International Trade and Policies for Genetically Modified Products

_____

International Trade and Policies for Genetically Modified Products

Edited by

Robert E. Evenson

Economic Growth Center,
Department of Economics,
Yale University, Connecticut, USA

and

Vittorio Santaniello

Dipartimento di Economia e Istituzioni,
Universita' degli Studi Roma 'Tor Vergata',
Rome, Italy

CABI Publishing

CABI Publishing is a division of CAB International

CABI Publishing
CAB International
Wallingford
Oxfordshire OX10 8DE
UK

Tel: +44 (0)1491 832111
Fax: +44 (0)1491 833508
E-mail: cabi@cabi.org
Website: www.cabi-publishing.org

CABI Publishing
875 Massachusetts Avenue
7th Floor
Cambridge, MA 02139
USA

Tel: +1 617 395 4056
Fax: +1 617 354 6875
E-mail: cabi-nao@cabi.org

Library of Congress Cataloging-in-Publication Data

Evenson, Robert E. (Robert Eugene), 1934-
  International trade and policies for genetically modified products / edited by Robert E. Evenson and
Vittorio Santaniello.
      p. cm.
  Includes bibliographical references and index.
  ISBN 0-85199-056-8 (alk. paper)
  1. Genetic engineering industry--Congresses. 2. International trade--Congresses. 3. Intellectual
property--Congresses. I. Santaniello, V. II. Title.
  HD9999.G452E84 2005
  382'.4566065--dc22

                                                                                    2004028436

Typeset by AMA DataSet Ltd, Preston, UK.
Printed by Cromwell Press Ltd, Trowbridge, UK.

Contents

Contributors

Blind, K., *Fraunhofer Institute for Systems and Innovation Research (ISI), Breslauer Str. 48, 76139 Karlsruhe, Germany.*

Brookes, G., *Brookes West, Jasmine House, Canterbury Road, Elham, Canterbury, Kent CT4 6UE, UK.*

Budd, G., *General Council, Grains Research and Development Corporation, Canberra, ACT 2604, Australia.*

Davey, K.A., *Department of Agricultural Economics, University of Saskatchewan, 51 Campus Drive, Saskatoon, Saskatchewan, Canada S7N 5A8.*

Eaton, D., *Agricultural Economics Research Institute (LEI), Wageningen University and Research Centre (WUR), PO Box 29703, 2502 LS, The Hague, The Netherlands.*

Frisvold, G.B., *Department of Agricultural and Resource Economics, University of Arizona, Tucson, AZ 85721, USA.*

Gray, R.S., *Department of Agricultural Economics, University of Saskatchewan, 51 Campus Drive, Saskatoon, Saskatchewan, Canada S7N 5A8.*

Haggui, F., *Department of Agricultural Economics, University of Saskatchewan, 51 Campus Drive, Saskatoon, Saskatchewan, Canada S7N 5A8.*

Kerr, W.A., *Department of Agricultural Economics, University of Saskatchewan, 51 Campus Drive, Saskatoon, Saskatchewan, Canada S7N 5A8.*

Khachatourians, G., *Department of Applied Microbiology and Food Science, University of Saskatchewan, 51 Campus Drive, Saskatoon, Saskatchewan, Canada S7N 5A8.*

King, J., *Economic Research Service (USDA), 1800 M Street NW, Rm 4195s, Washington, DC 20036, USA.*

Knudsen, O.K., *The World Bank, 1818 H Street, Washington, DC 20433, USA.*

Malla, S., *Department of Economics, University of Lethbridge, 4401 University Drive, Lethbridge, Alberta, Canada T1K 3M4.*

Menrad, K., *University of Applied Sciences of Weihenstephan, Science Centre Straubing, Schulgasse 18, D-94315 Straubing, Germany.*

Oehmke, J.F., *Department of Agricultural Economics, 317c AgH, Michigan State University, East Lansing, MI 48824, USA.*

Phillips, P.W.B., *Department of Agricultural Economics, University of Saskatchewan, 51 Campus Drive Saskatoon, Saskatchewan, Canada, S7N 5A8.*

Reeves, J.M., *Agricultural Research Department, Cotton Incorporated, 6399 Weston Parkway, Cary, NC 27513, USA.*

Sedjo, R.A., *Resources for the Future, 1616 P Street, NW Washington, DC 20036, USA.*

Scandizzo, P.L., *Facoltà de Economía Università, 'Tor Vergata', Via Columbia 2, 00133, Rome, Italy.*

Scandizzo, S., *Corporación Andina de Fomento, Torre CAF, Av. Luis Roche, Caracas, Venezuela.*

Schimmelpfennig, D., *Economic Research Service (USDA), 1800 M Street NW, Rm 4195s, Washington, DC 20036, USA.*

Schmitz, P.M., *Department of Agricultural Policy and Market Research, University of Giessen, Diezstr. 15, D-35390 Giessen, Germany.*

Smyth, S., *Department of Biotechnology, University of Saskatchewan, 51 Campus Drive, Saskatoon, Saskatchewan, Canada S7N 5A8.*

Tothova, M., *Department of Agricultural Economics, 317c AgH, Michigan State University, East Lansing, MI 48824, USA.*

Tran, K., *Department of Economics, University of Lethbridge, 4401 University Drive, Lethbridge, Alberta, Canada T1K 3M4.*

Van Tongeren, F., *Agricultural Economics Research Institute (LEI), Wageningen University and Research Centre (WUR), PO Box 29703, 2502 LS, The Hague, The Netherlands.*

Tronstad, R., *Department of Agricultural and Resource Economics, University of Arizona, Tucson, AZ 85721, USA.*

Weaver, R.D., *Department of Agricultural Economics, 207D Armsby Building, Pennsylvania State University, University Park, PA 16802, USA.*

Wesseler, J., *Environmental Economics and Natural Resources Group, Social Sciences Department, Wageningen University, Hoolandsweg 1, 6706KN, Wageningen, The Netherlands.*

Wronka, J., *Department of Agricultural Policy and Market Research, University of Giessen, Diezstr.15, D-35390 Giessen, Germany.*

Acknowledgements

The chapters in this volume were originally presented at the 8th International Conference of the International Consortium on Agricultural Biotechnology Research (ICABR), held at Ravello, Italy, on July 8–11, 2004. They have since been edited and revised.

The editors acknowledge sponsorship by the following:

- CEIS, University of Rome 'Tor Vergata', Rome, Italy.
- Economic Growth Center, Yale University, New Haven, Connecticut, USA.

Editors' Overview

This volume addresses international dimensions of agricultural biotechnology. The four chapters in Part I are primarily analytical. Part II includes two empirical trade chapters. Part III includes four chapters addressing spillover dimensions. Part IV includes three chapters addressing intellectual property rights (IPRs). Part V includes three chapters using applied general equilibrium trade models.

Part I includes four chapters with an analytic focus. Chapter 1 (Knudsen and Scandizzo) analyses interdependencies of risks and opportunities in biotechnology product development. The authors point out that the risks to the environment may appear to be small from the point of view of an individual project. However, as the number of projects increases, these risks can become substantial. They also note interdependencies on the opportunities side, where the number of projects and the number of adopters of GM products affect outcomes both negatively and positively. The application of the interdependency analysis is international in scope.

The authors show that a programme of n projects has a threshold value for acceptance that may be higher or lower than the threshold value for individual projects because of interdependencies. They argue that in project evaluation, two steps are required. First, the project should be appraised to consider its extended net present value, including the option to delay. Secondly, the project should be evaluated taking into account its impact on the overall programme.

Chapter 2 (Weaver and Wesseler) develops a model of monopoly R&D pricing with uncertain returns and irreversible costs and benefits. The monopoly power is associated with intellectual property rights (i.e. the right to exclude). In earlier work, Weaver and Kim showed that in the presence of alternative technologies, monopoly power is restricted. In this chapter, Weaver and Wesseler incorporate *ex ante* uncertainty and irreversibility of technology adoption in the model.

The authors find that both *ex ante* uncertainty and irreversibility reduce the willingness to pay of the technology purchaser and the pricing power of the technology provider. Irreversibility is illustrated by herbicide-tolerant soybeans where no-till equipment is required. These costs depend on the size of the farm. Irreversibility also applies to benefits, particularly externality benefits. These benefits affect costs and require a net irreversibility calculation. These effects can be both national and international in scope.

Chapter 3 (Tothova and Oehmke) emphasizes the direct trade in GM crops. They note that two distinct groupings of countries are emerging in formulations of regulatory regimes for GM foods. One group is the European Union (EU) 'club' where a combination of bans and restrictions on trade are being implemented. The second group is the North American 'club' with lower degrees of regulatory restrictions.

These two '*de facto*' clubs will form trading clubs for GM foods. The chapter analyses the nature of regulatory standards and conflicts with World Trade Organization (WTO) rules. The model is a Krugman-style monopolistic competition trade model. With full labelling and traceability, the two clubs are unlikely to agree on standardized regulatory systems. The likely outcome is for trade to occur between members of each club.

Chapter 4 (Scandizzo) develops an analysis of the role of labelling. The model uses a marketing technology exhibiting national increasing returns to scale.

Two cases are considered, no labelling and labelling. With no labelling, maximum national marketing economies of scale are achieved, but consumers cannot distinguish GM goods from non-GM goods. With labelling, marketing economies of scale will be lower. However, consumers will gain because they can now choose goods which give maximum utility. Conditions for which the loss of marketing scale economies outweighs the gains from consumer choice are stated in the chapter.

Part II includes two empirical trade chapters.

Chapter 5 (Sedjo) discusses developments in tree biotechnology and associated trade in transgenic wood and transgenic tree germplasm. The chapter notes that plantation forests based on 'improved' trees have grown rapidly in recent years and now account for 34% of wood products harvested in the world. Many plantation forests are based on selected 'exotic' species but increasingly conventionally bred (i.e. through sexual reproduction) and transgenics are being introduced to plantation forests. International trade in transgenic wood products is expected to be subject to fewer restrictions than trade in transgenic crops.

Chapter 6 (Smyth, Kerr and Davey) examines the international trading patterns for oilseed rape (canola), maize and soybeans. Comparisons are made for the period before the introduction of GM crops in Canada and the USA (1990–1996) and for the period after the introduction of GM crops (1997–2003).

The chapter reports a number of claims that the USA and Canada experienced large losses in export volume to the EU. A model of trade 'diversion' is developed in the chapter to show how exports in one market can be diverted to other markets.

The chapter reports that for GM rape seed exports and US GM maize seed exports, the EU market was lost, but that this was primarily trade diversion, since other countries imported more. The large US soybean export market also suffered EU market losses, but the EU remains the largest purchaser of US soybeans. (GM soybeans were approved for import by the EU prior to the 1998 moratorium.) China and Indonesia increased their imports of US soybeans to offset EU market losses.

Part III includes four chapters. The first addresses the question of coexistence of GM crops and organic crops. Two address spillovers. The fourth addresses regulations and innovation.

Chapter 7 (Brookes) provides an empirical evaluation of the coexistence issue. It notes that regulations regarding 'adventitious presence' of GM crops are important. (Current EU rules allow 0.9%, but the organic farm industry is asking for lower levels.) Chapter 7 notes that the total EU (15) area devoted to organic agriculture is about 3.5% of farmed area; 90% of this area is grassland. GM cropped area is only 0.35% of all cropped area in the EU.

The chapter reports acreages in organic production of oilseed rape, sugarbeet and maize, and projects future demand, and concludes that cropped area planted to organic crops will continue to be small (0.41% of cropped maize). An evaluation of coexistence in Spain where GM maize has been grown since 1998 leads the author to conclude that 'coexistence is possible without causing economic problems'.

Chapter 8 (Gray, Malla and Tran) evaluates research spillovers in the oilseed rape industry. The model is applied to firms in the rape industry in Canada. The modelling allows for research spillovers between firms in the industry and spillovers from public sector research programmes to private firms in the industry.

The study finds that both public basic and applied research had positive effects on private firm research productivity, profitability and social value output. In addition, spillovers within the industry were identified.

Chapter 9 (Schimmelpfennig and King) evaluates another type of spillover associated with mergers and acquisitions in the biotech industry. The study reported in the chapter was based on a new database containing 11,000 agricultural biotechnology patents granted between 1988 and 2002 by the US Patent Office. These patents were classified into nine technologies. The owners of the patent rights included small biotech firms, seed firms, chemical firms, multinationals and European firms, as well as public sector institutions.

The analysis concentrated on changes in patent ownership between the original assignee and the owner in 2002. A remarkable 65% of the patents changed hands. Thirty-five percent of the patents flowed from one category of firm to another.

Chapter 10 (Menrad and Blind) addresses the impact of regulation on innovation as measured by the development of new products in the food industry. The EU and US regulatory environments are compared.

Food industry firms have relatively high rates of new product introduction in both the EU and the USA (although reporting agencies do not consistently define 'new products'). The chapter evaluates studies of regulatory impacts on genetic engineering products, health-oriented functional foods and organic food products.

Part IV includes three chapters addressing issues associated with IPRs. These are particularly relevant for developing countries in view of the Trade Related Aspects of Intellectual Property Rights (TRIPS) provisions of the WTO.

Chapter 11 (Eaton and van Tongeren) compares alternative IPR systems for agriculture. They base their analysis on in-depth interviews with 12 international plant breeding companies.

The authors develop a framework for analysing the effects of IPRs that explicitly incorporates transaction costs associated with acquiring and enforcing IPRs. They find that the effectiveness of plant variety protection (breeders' rights) was identified as an IPR sensitive to transaction costs. The design and implementation of IPR systems is an important factor in effectiveness.

Chapter 12 (Smyth, Khachatourians and Phillips) raises issues associated with gene flow in the case of plant-made pharmaceuticals (PMPs). PMPs have considerable economic potential. The prevailing use of mammalian cells to produce pharmaceuticals could be replaced by PMP technology. Several crop species have been in PMP trials since 1992.

The production of PMPs is especially sensitive to gene flow problems. Regulatory agencies have imposed penalties on managers of PMP trials in several cases. Since the acreage of cropland devoted to PMP production is likely to be low, the authors propose several regulatory devices to prevent gene flow.

Chapter 13 (Budd) addresses the use of genetic use restriction technologies (GURTs) in agriculture. Actually, one of the technical solutions to the gene flow problem discussed in Chapter 12 would be the use of GURT technology.

When GURT technology was first introduced several years ago, it was widely criticized (the terminator gene). The CGIAR institutes stated that they would not use GURTs. There was a fear that farmers would lose the right to save their own seed. The GURT effectively imposes seed sterility. GURTs were condemned by a number of organizations.

Chapter 13 shows these fears to be unfounded. As long as some competition exists in the seed industry, even if provided by public sector institutions, GURTs are not an effective form of IPR. The reason is quite simple. The installation of a GURT in a plant only endows the plant with a sterile seed system. This reduces the value of the seed to the farmer, but does not provide any real benefits to the seed buyer (as do GM crops or hybrid crops). Thus, as long as a total monopoly on seed supply is in place, GURTs will be installed only when the crop variety has a large inherent advantage over its nearest competitor. Since most countries do not have seed monopolies (almost all countries have public experiment station systems), GURTs cannot be used as IPRs.

Part V includes three chapters that use applied (or computable) general equilibrium (AGE) models to evaluate policy impacts.

Chapter 14 (Wronka and Schmitz) uses a multicommodity multiregion model to evaluate alternative consequences of GM crop production in the USA, Brazil and China under three different assumptions regarding EU response to GM crop production. In the first scenario, the EU bans imports of GM crops. In the second, the EU imports GM crops. In the third, the EU allows GM crop production.

In the first scenario, Brazil, China and the USA gain welfare while the EU loses welfare. Producers in the EU gain, but consumers lose. In the second scenario, Brazil, China and the USA gain. In the EU, producers lose, but consumers gain more so that welfare improves in the EU. In the third scenario, all countries and regions gain.

Chapter 15 (Haggui, Phillips and Gray) analyses the impact of Monsanto's decision to cease the development of its Roundup Ready varieties of GM wheat on global food security (GFS). The authors utilized a GFS model for GM wheat (developed in the chapter). The model includes adoption rates in the presence or absence of opposition to GM crops.

The authors report that widespread opposition to GM wheat leads to a significant increase in the price of wheat (17%) and reduced consumption, production and stocks. Limited opposition (e.g. Japan and Korea) would have relatively small effects. Widespread opposition would exacerbate global food security.

Chapter 16 (Frisvold, Tronstad and Reeves) analyses the international impacts of Bt cotton in a three-region model with an endogenous (model-determined) output price. The study finds that Bt cotton adoption reduced world cotton prices. Global benefits in 2001 were US$900 million, with producers getting US$600 million. Consumers received US$150 million, and seed suppliers received US$150 million. The adoption of Bt cotton in the USA and China imposed losses of US$350 million on producers who did not have access to Bt cotton technology. (For 2003 these estimates would be roughly double those reported.)

1 Biotechnology Risks and Project Interdependence

Odin K. Knudsen[1] and Pasquale L. Scandizzo[2]

[1]The World Bank, Washington, DC, USA; [2]The University of Rome 'Tor Vergata', Italy

Introduction

Biotechnology projects are characterized by the fact that their risks to the environment seem to appear minuscule from the point of view of an individual project, but the potential for damage grows with the number of projects. This characteristic transpires from many of the studies on this issue, but, at the same time, the scientific community does not appear to realize fully the significance of the greater risks that may be associated with this form of interdependence. A case in point is given by a major report produced by a formal Committee of the US National Research Council (2000), which addressed several risks of genetically modified (GM) crops, all virtually compounded by interdependence among projects.

Limiting our remarks to the ecological risks, in particular, the Committee considered three major impacts of pest-protected plants: (i) effects on non-target species; (ii) effects of gene flow; and (iii) evolution of pest resistance to pest-protected plants. On the first issue, the Committee concluded that, 'Both conventional and transgenic pest-protected crops could have effects on non-target species, but these potential effects are generally considered to be smaller than the effects of broad-spectrum synthetic insecticides. Therefore, the use of pest-protected crops could lead to greater biodiversity in agro-ecosystems where they replace the use of those insecticides'. Because replacement can only occur gradually as the new technology replaces the older one, this diagnosis recognizes the importance of interdependence between the degree of implementation of projects based on the use of the transgenic crops and those still based on the use of insecticides. Furthermore, as transgenic crops spread to more species and insecticides are progressively replaced, interdependence between projects based on different transgenic species would also have to be recognized.

As for the effects on non-target organisms, the committee recognized the risks involved in widespread dispersal of pollen containing insecticidal genes. These risks arise when large regions are planted with wind-pollinated biotechnology (Bt) plants, as a consequence of the implementation of several projects including these techniques. Because Bt pollen dispersal tends to grow geometrically with the area planted, interdependent risk to non-target species across the area utilized by different Bt projects, sharing this technique, itself grows more than proportionally with the number of projects.

Similar remarks can be made regarding the other two ecological concerns: gene flow and pest resistance. For the first, the Committee concluded that the transfer of either conventionally bred or transgenic resistance traits to weedy relatives potentially could exacerbate weed problems, but such problems have not been observed or adequately studied. For pest resistance, the Committee stated that it 'can have a number of potential environmental and health impacts'. It also recommended that resistance management strategies be developed for specific pest-protected plants, and implementation of such practices (i.e. through new, related projects) be encouraged.

Against this prudent judgement by the Committee, however, stand several concerns of the public and the governments involved in the controversy over the so-called genetically modified organisms (GMOs). In this debate, the European Union (EU) and Japan have appeared to be much more concerned about possible dangers to the environment and consumers' health than American authorities and scientists. Accordingly, in a highly controversial decision, in June 2002, the EU's environment ministers imposed a moratorium on new marketing approvals for GMOs. Furthermore, ministers from Denmark, France, Greece, Italy and Luxembourg took a common stance to deny new applications to market GM seeds, plants or foodstuffs until a new regulatory regime is put in place. In another example, at the same time that the EU was taking such a position, the Japanese Ministry of Agriculture, Forestry and Fisheries (MAFF) stated that it would deny approval to Bt crops in Japan until safety regulations on GM crops are reformulated in a more stringent and satisfactory way. This decision came in the wake of a Cornell University study, which was considered, but discounted by the US NRS Committee. The study showed that pollen from insect-resistant Bt (*Bacillus thuringiensis*) maize is potentially toxic to monarch butterfly larvae.

The fact that biotechnology products appear to stir so many controversial and diverse reactions largely depends on the perception of public risk. Whilst the traditional conception of risk assumed its appreciation to depend on the preferences (the degree of 'risk aversion') of the decision maker, recent developments of theory and measurement have gradually developed a different perspective. Many of these new approaches (Slovick, 1991; Beck, 1992; Nelkin, 1995) conform to the view that uncertainty and risk are, respectively, the public and the private good side of the same medal. Uncertainty can thus be seen as an externality, being internalized in different ways by the stakeholders involved. As a consequence, risk perception depends on the degree of empowerment of those who internalize the uncertainty. As such, it is influenced by social relations, feelings of powerlessness and lack of trust in the face of fear of large-scale accidents, individual harm and

deterioration of the quality of life. Whilst traditionally risk was seen as a problem regarding only the individuals directly concerned with danger or the prospect of a loss, today there is an increasing sense that risk has become a 'social good' and that social standards are required to deal with the emergence of a 'risk society'. In turn, this is characterized by increasing interdependence of scientific and economic endeavours, widespread implications of seemingly local or isolated events and controversies that undermine the basis for calculating risk and insurance (Beck, 1992).

One of the major reasons why biotechnology appears threatening is the fact that its actions 'meddle with nature' in ways that are unexperimented in their broader and possibly irreversible effects. The build-up of pest resistance through gene drift is an example of irreversible damage, that could spread far beyond the GMOs that originate the drift. An even more dangerous possibility is the creation of harmful microorganisms. Both examples suggest the possibility that a catastrophic change in the ecological balance may be created by Bt projects that, seen individually, may appear to have limited scope and to carry non-significant, local risks.

In addition to the fear that biotechnology may be an instrument of danger by diffusion, Bt products suggest further risks due to the fact that they give rise to externalities that are difficult to internalize. While open-market prices have traditionally provided signals for production and distribution that have resulted in efficient commodity production, Bt products are far more difficult to coordinate across buyers and sellers than traditional products. Vertical coordination will be increasingly difficult as the new Bt products on the market target the needs of the end-user or consumer. Such is the case, for example, of foods with altered nutritional qualities, crops with improved processing characteristics and plants that produce specialty chemicals or pharmaceuticals. Producing and marketing these commodities implies interdependencies across the value chain that require greater control and more formal information than the existing open-market system. These interdependencies arise from the fact that succeeding additions to the value changes such as genetics, production practices, harvesting, handling, storage and processing are closely correlated with one another. Desired traits for end-users, in fact, can be preserved only if the different stages of value added move together according to certifiable protocols and testing that may validate product content. As coordination increases, correlation of risks and value across stages becomes larger, and both risks and additional value are shared among the market participants.

Horizontal coordination among Bt projects is also problematic. For example, virtually all ecological risks of biotechnology appear to depend on implementation of several projects that share common risk factors (Bt pollen, gene spillover and pest evolution) and display effects, as a consequence, positively correlated with one another, or negatively correlated with projects that, by using traditional technologies, are based on different risk factors (e.g. chemical insecticides versus pest-resistant species).

In addition to interdependent risks, biotechnology projects are also characterized by interdependent opportunities. One example is the improvement of environmental conditions due to the reduction of chemical pesticides, which is positively correlated with improved nutrition that high yield Bt crops could provide and with Bt applications in health and hygiene. Another example is the case of Bt applications in food processing. This is positively correlated with some of the new Bt crop varieties, the diagnostics for plant and animal diseases, and vaccines against animal diseases.

While the case for the existence of interdependent risks and opportunities in Bt projects can be plainly made, it is not apparent that these interdependencies are appreciated by governments and the private sector. Because projects are appraised on the basis of expected net present values (NPVs), interdependencies on other projects are neglected, even in the case where the projects are part of a programme which should provide for integration and coordination. The main effect of the neglect of project interdependencies can be easily predicted: Bt risks and opportunities will generally be ignored at project level because they do not appear large enough to bother. Every secondary effect of a project on the environment, health or production, in fact, will appear negligible if the project is not unduly large. Failure to recognize that these effects (which may be good or bad) may be significant, because they tend to be compounded by the fact that many interdependent projects are implemented, may result in systematic over- or underinvestments and deprive project evaluation methodology of much of its power of project selection.

In this chapter, we look at the problem of evaluating Bt interdependent projects in the framework of the theory of real options. First we present the case for the existence of stochastic interdependencies in Bt products. Next we develop a simple model of project selection under dynamic uncertainty, where failure to consider cross-project correlations results in suboptimal investment. Finally, we discuss the results and their policy implications.

The Interdependent Nature of Biotechnology Risks

Bt products present an unusual challenge to evaluation methodologies, because of the broader and unpredictable process of transformation in which they seem embedded. As this process unfolds, scientific uncertainty appears to grow and unforeseen possibilities to arise. The key characteristic of this process can be characterized as one of scientific uncertainty (Wesseler, 2002), where both benefits and costs from Bt projects are stochastic and their relationship fluctuates with uncertain pay-offs and possible losses.

More specifically, after a first phase where Bt products were mainly tailored at increasing production efficiency, by reducing costs and increasing productivity, a new phase is now emerging where enhanced consumption characteristics are sought and developed. In both cases, the technological links among the different products and processes of production appear to be growing in number and strength. Cross-fertilization and synergies among products is already high in the production area, but they promise to be much higher in the realm of product characteristics, where enhanced qualities for consumer preferences are sought across the spectrum of species and genes.

Table 1.1 shows that the first stage of Bt innovations had to do with improving the process of production. Insect resistance and herbicide tolerance were common characteristics developed through similar techniques and different crops. Correlations among methods, results and interactions in the field depended principally on the fact that the new products shared similar processes of production. In this context, risks to the environment mainly originate from possible spillovers from common technical factors.

Table 1.1. Genetically engineered products: transgenic agricultural products approved for unregulated release as of May 1999.

Products	Value-enhanced traits
Crop	
Beet	Herbicide tolerance
Carnation	Altered flower colour
Chicory	Herbicide tolerance
Maize	Herbicide tolerance
	Insect resistance (Bt)
Cotton	Herbicide tolerance
	Insect resistance (Bt)
Flax	Herbicide tolerance
	Insect resistance (Bt)
Papaya	Virus resistance
Potato	Insect resistance (Bt)
Rapeseed	Herbicide tolerance
	High lauric acid oil
Rice	Herbicide tolerance
Soybean	Herbicide tolerance
	High oleic acid oil
Squash	Virus resistance
Tomato	Delayed ripening
Non-crop	
Chymosin	Enzyme used in cheese production; produced in bacteria
RBST	Bovine growth hormone; produced in bacteria

Source: Economic Issues in Agricultural Biotechnology/AIB-762; Economic Research Service/USDA.

Table 1.2 shows how the 'new' wave of Bt products is taking shape. In addition to the increase and enhancement of 'input traits' that were common to the first generation of GMOs, the new developments expand the scope of Bt technology to various 'output traits'. These traits are all characterized by enhancements of product appeal to consumers via the introduction of desirable characteristics. While the risks arising from the techniques of genetic manipulation remain the same, broadening the range of product characteristics is likely to result in a higher level of interdependence among products and projects. Synergies across enhanced products are likely to arise because of complementarities between characteristics, such as different nutrients in grain and rice. Substitution relationships may also build correlations among products, while vertical interdependence along the value chain will also give rise to covariant risks and opportunities.

A powerful element of linkage across projects is also due to the fact that biotechnology, through the increased spread of new products, appears to challenge the traditional protocols for production and consumption. Genetic manipulation thus raises fundamental ethical questions and deep fears that bind Bt products beyond what may be the present evidence for existing links among them. In other words, in addition to the characteristics developed, Bt manipulation generates a by-product of 'shadow'

Table 1.2. Genetically engineered products: examples of transgenic products 'in the pipeline'.

Input traits
- Introduction of herbicide tolerance into sugarbeet, wheat, lucerne, sugarcane, potatoes, forestry products, specialty fruits and vegetables.
- Production of insect resistance in tomato, sugarcane, soybeans, rapeseed, groundnuts, aubergines and poplar; includes using other Bt toxins with different specificities and developing other toxins that could alleviate the problems associated with development of resistance to Bt.
- Introduction of disease resistance (to viruses, fungi and bacteria) in maize, potatoes and a variety of fruits and vegetables.
- Introduction of genes for other agronomic traits, including drought tolerance, frost tolerance, enhanced photosynthesis, more efficient use of nitrogen and increased yield.
- Increasing use of 'stacked' traits (herbicide tolerance and Bt resistance in one plant, for example).

Output traits
- Feed quality, food quality, value-added traits, specialty chemical production.
- Traits affecting quality of animal feed.
- Low phytate maize.
- Soybeans and maize with altered protein or oil levels (nutritionally dense).
- Traits affecting food quality for human nutrition (nutraceuticals):
    Oilseed rape and soybeans producing oils high in stearate or low in saturated fats.
    Oilseed rape with high β-carotene (antioxidant) content.
    Tomatoes with elevated lycopene levels (anti-cancer agent).
    Grains with optimized amino acid content.
    Rice with elevated iron levels.
    Increased vitamin content.
    Production of 'low-calorie sugar' (indigestible fructans) in sugarbeets.
    Increased sugar levels in maize and strawberries, for enhanced flavour.

Traits that affect processing
- Coloured cotton.
- Cotton with improved fibre properties.
- High-solid tomatoes and potatoes.
- Delayed ripening fruit and vegetables, such as melon, strawberries and raspberries.
- Altered gluten levels in wheat to alter baking quality.
- Naturally decaffeinated coffee.

Production of specialty chemicals (plants as bioreactors)
- Production of pharmaceuticals, antibodies, vaccines and industrial chemicals in transgenic plants, e.g. include diarrhoea vaccines in bananas, blood proteins in potatoes, rabies vaccine in maize and monoclonal antibodies in maize.
- Transgenic livestock
    Pharmaceuticals produced in milk in cows, pigs, or sheep; examples include antithrombin III (a blood anticoagulant, currently in phase III clinical trials), α-1-antitrypsin (used to treat cystic fibrosis).
    α-Lactalbumin (a human milk protein to use as a nutritional supplement).
    Livestock with more rapid growth, less fat, disease resistance; more long term.

Sources: Information Systems for Biotechnology website at Virginia Polytechnic Institute and State University (www.isb.vt.edu); APHIS Agricultural Biotechnology website (http://www.aphis.usda.gov/biotech/); Biotechnology Industry Organization website (www.bio.org); Monsanto website (www.monsanto.com); OECD BioTrack Online website (www.oecd.org//ehs/Service.htm).

characteristics, that destabilize benefits and costs across products and projects.

It is important to realize that the form of stochastic interdependence brought about by the above combination of factors tends to increase the costs and delay the adoption of the new technologies, precisely when synergies, rather than negative relationships, arise across projects. A higher level of correlation among benefits of Bt projects, in fact, may increase project variance by building up covariant risk, just as much as does correlation among costs. As a consequence, projects are more vulnerable to repeated shocks, since these tend to affect all positively related projects in a similar way. Stakeholders are also likely to suffer from this process, since both consumers and producers across products and along the value chain share the risk that a random shock may cause a widespread loss to all interest bearers.

The Model

Consider a class of biotechnology projects whose stream of benefits are uncertain, for example the release of transgenic crops. We assume that the benefit stream from the projects taken together is governed by a zero drift, Wiener process of the linear variety:

$$dy = \sigma d\zeta \qquad (1)$$

where $\sigma > 0$ and $d\zeta$ is a random variable with zero expectation and variance equal to dt (i.e. to the smallest variation of time considered). Because linear processes are additive, they reproduce themselves. Thus, the stochastic process in Equation 1 can be expressed as the sum of the stochastic processes underlying each project ($i = 1, 2, n$):

$$dy = \sum_{i=1}^{n} dy_i = \sum_{i=1}^{n} \sum_{j=1}^{n} \sigma_{ij} d\zeta_i \qquad (2)$$

where $\sigma_{ii} > 0$, and $\sigma_{ij} \geq 0$ or < 0, $Ed\zeta_i = 0$, $Ed\zeta_i d\zeta_j = dt$, for $i, j = 1, 2, \ldots, n$. Equation 2 says that, once adopted, each project benefit stream affects all other project streams. On the other hand, if the project is not adopted (and until is adopted), it is irrelevant for the others.

Given this structure, if the projects are evaluated as a programme, the option to undertake the programme, $F(y)$, can be characterized through dynamic programming, using the condition of optimum in continuation time (Bellmann's equation):

$$rF(y) = EdF(y) \qquad (3)$$

where r is an appropriate discount rate. Applying Ito's lemma to Equation 3 yields:

$$rF(y) = \frac{1}{2} F'' \sigma^2, \quad \text{where} \quad \sigma^2 = \sum_{i=1}^{n} \sum_{j=1}^{n} \sigma_{ij} \qquad (4)$$

As a general solution of the partial differential equation in (4), we try the functional form: $F(y) = Ae^{\beta y}$ which, upon substitution into Equation 4, yields the characteristic equation:

$$r - (\beta^2 \sigma^2)/2 \qquad (5)$$

whose solution is given by two roots of opposite sign:

$$\beta = \pm \frac{\sqrt{2r}}{\sigma} \qquad (6)$$

The value matching condition for the decision to adopt the programme may be written as follows:

$$A \exp\left(\frac{\sqrt{2r}}{\sigma}\right) = \frac{y}{r} - I, \quad I = \sum_{i=1}^{n} I_i \qquad (7)$$

where the negative root of Equation 5 has been discarded because it does not concern the option to enter (since it is a 'call' option, requiring the value of the option to increase with the value of the underlying). The right hand side of Equation 7 represents the cash flow promised by the entire programme as a consequence of simultaneously committing resources I to all projects.

The smooth pasting condition, on the other hand, can be simply written as:

$$A \exp\left(\frac{\sqrt{2r}}{\sigma}\right) = \frac{\sigma}{r\sqrt{2r}} \qquad (8)$$

Substituting Equation 8 into Equation 7 and solving for the cash flow y, yields the threshold of acceptance y^* for the programme:

$$\frac{y^*}{r} = I + \frac{\sigma}{r\sqrt{2r}} \qquad (9)$$

Consider now the stochastic process underlying a single project. This can be written as:

$$dy_i = \sum_{j=1}^{n} \sigma_{ij} d\zeta_i d\zeta_j \quad i = 1, 2, \ldots, n$$

for some $d\zeta_j \neq 0$ and $dy_i = \sigma_{ii} d\zeta_i$,

if $d\zeta_j = 0$ for all $j \neq i$ (10)

so that the option value is defined as:

$$F(y_i) = A_i \exp\left(\frac{\sqrt{2r}}{\sqrt{\sum_{j=1}^{n} \sigma_{ij}}}\right) y_i \qquad (11)$$

where $\sigma_{ij} = 0$ for all j such that the correspondent project is not in operation.

Following the same procedure used for the programme, the threshold of acceptance for the ith project will be:

$$\frac{y_i^*}{r} = I_i + \frac{\sqrt{\sum_{i=1}^{n} \sigma_{ij}}}{r\sqrt{2r}}; \quad i = 1, 2, \ldots, n \qquad (12)$$

where again $\sigma_{ij} = 0$ for all j such that the correspondent project is not in operation. In other words, the option value of the ith project and its threshold of acceptance will include the covariances only to the extent that the underlying process is activated through a correspondent project.

Proposition 1

Given a programme formed by n projects, a separate appraisal of each project will cause overall under- or overinvestment according to whether the projects in the programme are positively or negatively correlated with one another.

Proof

The Bellmann equation of dynamic programming prescribes (Dixit and Pindyck, 1994, p. 105) that for optimality, the following condition must be met by the value of any asset held by the decision maker:

$$rF(y, t) = \max_u \left\{ \pi(y, u, t) + \frac{1}{dt} E[dF] \right\} \qquad (13)$$

where π are the dividends paid by the assets and u is a control variable for these dividends.

Because holding the option to adopt the project does not pay any dividend, Equation 13 can be written in terms of Equation 3 and solved to give:

$$F(y) = A \exp\left(\frac{\sqrt{2r}}{\sigma}\right) \qquad (14)$$

Since this is the value that verifies the Bellmann condition, any other value is not a maximum, i.e.

$$A \exp\left(\frac{\sqrt{2r}}{\sigma}\right) \geq \sum_{i=1}^{n} A_i \exp\left(\frac{\sqrt{2r}}{\sigma_i}\right) \qquad (15)$$

Thus, not considering the correlations among the stochastic processes concerned has the effect of giving a value to the programme option which is below or equal to the sum of the option values for the individual projects. In other words, the decision makers concerned with individual projects will fail to maximize the value of the programme.

Furthermore, by summing the project thresholds of acceptance in Equation 10, under the assumption that all other projects are not adopted, and subtracting the sum from the programme threshold in Equation 9, we find:

$$\frac{y^*}{r} - \sum_{i=1}^{n} \frac{y_i^*}{r} = \frac{\sigma - \sqrt{\sum_{i=1}^{n} \sigma_{ij}}}{r\sqrt{2r}} \qquad (16)$$

Q.E.D.

Thus, if each project is appraised in isolation, the threshold value for the sum of the projects will differ from the sum of the thresholds of an amount equal to the sum of the covariances between each pair of projects. In other words, if each project is appraised independently, resources to the programme will be committed too soon (overinvestment) or too late (underinvestment) according to whether the sum of the covariances is positive or negative.

Comment

The difference between the thresholds arises because the extended NPV (ENPV) (i.e. the NPV minus the waiting option) of the programme is different from the corresponding project measure. The difference depends on the value of the option, which is affected by the covariances in the case of the programme, but not for the individual projects. In fact, deriving the value of the constant A, by combining Equations 8 and 9, we find, as the value of the programme option:

$$Ae^{\beta y} = \frac{\sigma}{r\sqrt{2r}} \exp\left[\left\{ \frac{\sqrt{2r}}{\sigma} \left(\frac{y}{r} - I \right) \right\} - 1 \right] \qquad (17)$$

This value clearly differs from the sum of the values of the project options, because, unlike the latter, it takes into account the covariances. In the case of many projects, the traditional prescription was to implement them on the basis of whether NPV ≥ 0. With option value linking the projects through their covariances, however, this simple rule breaks down. The adoption of each project, in fact, potentially affects all subsequent decisions and is itself affected by the possibility of adoption of other related projects.

Proposition 2

The optimal social rule to decide whether to adopt a project that is part of a programme is to consider for each project an ENPV including the option to wait, under the assumption that all other projects are undertaken.

Proof

Consider the value matching condition for the ith project, under the assumption that all other projects are adopted. The value matching condition is given by Equation 12, computed for all non-zero elements. However, in this case, Equation 12 yields a zero value and the social optimum from programme adoption coincides with the sum of the values obtained by the individual entrepreneurs appraising each project in isolation.

Proposition 3

For a programme formed by a set of projects whose cash flows are stochastically correlated, an optimal set of taxes (subsidies) can be found such that: (i) the sum of the thresholds of adoption for each project equals the threshold of the sum (i.e. the whole programme); and (ii) the difference of the entry points of each project is as low as possible.

Proof

Consider the case of n projects each to be adopted by a different agent, and assume that the underlying stochastic processes are correlated. If each of the n agents decides in isolation, disregarding the correlation with the other project, the difference between the programme and the projects threshold level would be:

$$\frac{y^*}{r} - \sum_{i=1}^{n} \frac{y_i^*}{r} = \frac{\sigma - \sqrt{\sum_{i=1}^{n} \sigma_i^2}}{r\sqrt{2r}} \qquad (18)$$

In order to conform individual behaviour to social efficiency, a tax corresponding to this difference has to be distributed among the n agents in such a way that projects will enter only if $y_1 + y_2 + \cdots y_n \geq y^*$. What is the best distribution? Consider the tax rate τ_i, such that: $y_i(\tau_i) = y_i(1 - \tau_i)$.

$$\tau_i = 1 - \exp\left\{\frac{\sqrt{2r}}{\sigma_i} - \frac{\sqrt{2r}}{\alpha_i \sigma}\right\} y_i \qquad (19)$$

where α_i is a weight such as $\sum_{i=1}^{n} \alpha_i = 1$.

It is easy to see that applying this specific tax rate to each project net cash flow is equivalent to transform the corresponding value matching condition into:

$$A_i \exp\left\{\frac{\sqrt{2r}}{\alpha_i \sigma}\right\} y_i = \frac{y_i}{r} - I_i \qquad (20)$$

so that the threshold of entrance will be:

$$\frac{y_i^*}{r} = I_i + \frac{\alpha_i \sigma}{r\sqrt{2r}} \qquad (21)$$

Proposition 4

In order to ensure that the projects are implemented at the same time as the programme, projects weights α_i in the optimal tax should be equal to the ratios between expected net benefits of each project and expected net benefits from the overall programme.

Proof

Deriving the value of the constant from Equation 20 for $y_i = y_i^*$, yields:

$$A_i \exp\left\{\frac{\sqrt{2r}}{\alpha_i \sigma}\right\} y_i = \frac{\alpha_i \sigma}{r\sqrt{2r}} \exp\left[\frac{\sqrt{2r}}{\alpha_i \sigma}\right](y_i - y_i^*) \qquad (22)$$

By definition, the value of project option at any time is equal to its expected discounted value at adoption:

$$A_i \exp\left\{\frac{\sqrt{2r}}{\alpha_i \sigma}\right\} y_i = E e^{-rt_{ii}} \frac{\alpha_i \sigma}{r\sqrt{2r}} \qquad (23)$$

From (22) and (23) it thus follows that:

$$E e^{-rt_i} = \exp\left[\frac{\sqrt{2r}}{\alpha_i \sigma}\right](y_i - y_i^*) \qquad (24)$$

Thus, equal timing is ensured if:

$$\left[\frac{\sqrt{2r}}{\alpha_i \sigma}\right](y_i - y_i^*) = \left[\frac{\sqrt{2r}}{\sigma}\right](y_i - y^*)$$

for all $i = 1, 2, \ldots, n$ 　　　　(25)

Solving for α_i, after substituting the expressions for y_i^* and y obtained respectively in Equations 21 and 9, we find:

$$\alpha_i = \frac{y_i - I_i}{y - I}$$ 　　　　(26)

Q.E.D.

Comment

If project uncertainty generates a social risk, by the precautionary principle (Knudsen and Scandizzo, 2005), we should seek the distribution that minimizes the risk that any of the n agents will adopt the project before the programme. The programme is thus to set the shares α_i in such a way that the projects are all adopted at the same time and that this will also coincide with the adoption timing of the programme. In order to do that, project weights in the taxation schedule should be continuously adjusted to reflect the ratios of net benefits between the individual project and the programme. The higher the share of total net benefits claimed by the individual project, the proportionally higher the tax rate to be charged to the project to ensure that the programme is not started until all programme components are mature.

Proposition 5

In order to ensure that programme components may be implemented in a way consistent with social optimality, with tax rates constant over time, project weights α_i in the optimal tax should be a linear function of the difference between the average and the individual project size.

Proof

In order to find a set of constant weights, we can minimize the difference between the entry points, by setting the α_i accordingly.

More specifically, we can find the values of α_k, $k = 1, 2 \ldots n$; such that:

$$\min_{\alpha_{xk}} L = \sum_{i=1}^{n}\sum_{j=1}^{n}\left(I_i - I_j + \frac{(\alpha_i - \alpha_j)\sigma}{r\sqrt{2r}}\right)^2$$ 　　　(27)

subject to: $\sum_{k=1}^{n} \alpha_k = 1$.

As can be readily seen, the solution to Equation 27 is:

$$\alpha_k = \frac{1}{n} + \left(\frac{I}{n} - I_k\right)\frac{\sigma}{r\sqrt{2r}}$$ 　　　(28)

The share of the burden for each project can thus be divided into two parts: (i) a uniform share equal for all projects; and (ii) a correction proportional to the difference between the average and the specific project size. Substituting into Equation 19, we find:

$$t_k = 1 - \exp\left\{\left[\frac{1}{\sigma_k} - \frac{1}{\frac{\sigma}{n} + \left(\frac{I}{n} - I_k\right)}\right](\sqrt{2r})y_k\right\}$$

for $k = 1, 2, \ldots, n$ 　　　(29)

Expressions 28 and 29 imply that each project should be charged with a tax that is an increasing function of its standard deviation (SD) and a decreasing function of the programme SD plus the difference between the average project investment and the investment of the specific project. In other words, projects with lower entrance points should be charged while projects with higher entrance points should be subsidized.

Q.E.D.

Implications for Investment Policies and Project Design

The field of biotechnology appears to be one where risks and opportunities are related to one another by sharing common factors of determination and interaction. These factors tend to be recognized for research and production, but are generally ignored at the project level. A project can be conceived as a planned intervention on the environment, with several effects, which may be positive and negative. These effects, in the case of biotechnology, tend to be stochastic, irreversible and correlated across projects sharing common Bt factors and components. In particular, it is important to emphasize the fact that they are not only negative effects, such as the negative externalities that are feared *vis a vis* the environment. Correlation of underlying stochastic processes across projects may arise for a variety of circumstances, including the well known case of correlation between prices and output.

The main implication of the existence and the significance of these correlations is that, in order to avoid the so-called 'macro-fallacy', the decision to adopt an individual project cannot be taken without considering, at the same time, the opportunity to adopt the group of projects, i.e. the programme, to which the project belongs explicitly as a consequence of policy choices, or implicitly by virtue of the existing correlations. What do these correlations do and how do they affect project stance in the selection process? Once considered through the lenses of real option theory, the answer is surprisingly simple: correlations increase, when they are positive, and reduce, when they are negative, programme variance. Hence, they affect the value of the option to delay programme adoption and, indirectly, the desirability to adopt any individual project in the programme. Of course, covariant risks have always been a problem in agriculture, since, for a given area, many shocks (weather, pests and health problems) are likely to influence most enterprises located in the area. For Bt projects, however, we are arguing for a higher dimension of covariant risk, since covariances would arise by the very innovations that the projects carry. For example, the risk that an individual project based on insecticidal genes significantly affects pest resistance may be small, but covariant risk across many projects and a larger area may be significant.

In order to take into account correlations in project evaluation, therefore, two separate steps are called for. First, the programme should be appraised to compute its ENPV, which includes the value of the option to delay. Secondly, each project should be evaluated taking into account its impact on the overall programme. Thus, no project should be undertaken if the correspondent programme is not cleared for action. At the same time, even if the programme is considered worth implementing immediately, an individual project may still be rejected, at least temporarily, if the other projects of the programme are not implemented simultaneously as well.

A second implication concerns project design. Given the significant correlations that Bt projects are likely to display across territories, species and genes, no single intervention may be considered 'small', if the probability of implementation of other projects that impact or drive on the same factors is sufficiently high. Because the correlations tend to act as effects external to the projects, larger and more homogenous projects, which internalize them, appear in principle to be more efficient than smaller projects that will be more likely to ignore these externalities. The rural development project model, in particular, seems badly suited to accommodate Bt innovations. Its diversified structure and the relative small size of its components, in fact, could increase the danger that the volatility of its effects be compounded by the correlations with other projects and/or project components.

A third implication applies to the use of the so-called precautionary principle. As we have shown elsewhere (Knudsen and Scandizzo, 2001), this principle of prudence, which is increasingly being built into the evaluation process in many countries and in the EU in particular, can be interpreted as the shifting of the burden of proof to the demonstration that, when undertaking a Bt project, a significant risk does not exist. However, applying this principle to the individual project would clearly miss the point, if the risk arising from the implementation of projects with covariant net benefits were not taken into consideration. At the programme level, the precautionary principle will be more stringent than at the project level, if covariant risk is greater than individual risks (negative correlation across projects), and less stringent otherwise. Also, programmes can be designed in such a way that the principle is satisfied because of the composition of the programme and the structure of the projects involved.

References

Beck, U. (1992) *Risk Society: Towards a New Modernity.* Sage Publications, London.

Dixit, A.K. and Pindyck, R. (1994) *Investment Under Uncertainty.* Princeton University Press, Princeton, New Jersey.

Knudsen, O. and Scandizzo, P.L. (2001) *The Precautionary Principle and the Social Standard.*

World Bank SSD Working Paper. World Bank, Rome.

Knudsen, O. and Scandizzo, P.L. (2005) Bringing social standards in project evaluation under dynamic uncertainty. *Risk Analysis* (in press).

Nelkin, D. (1995) Science controversies. In: Jasanoff, S., Markle, G.E., Persen, J.C. and Pinch, T. (eds) *Handbook of Science and Technology Studies.* Sage Publications, Thousand Oaks, California, p. 445.

Slovic, P. (1991). Beyond numbers: a broader perspective on risk perception and risk communication. In: Mayo, D.G. and Hollander, R. (eds) *Acceptable Evidence: Science and Values in Risk Management.* Oxford University Press, New York, pp. 48–65.

United Nations Development Program (1999) *Human Development Report 1999.* Oxford University Press, New York, p. 36.

US National Research Council (2000) *Genetically Modified Pest-Protected Plants – Science and Regulation.* US National Research Council.

Wesseler, J. (2002) Assessing the risk of transgenic crops – the role of scientific belief systems. In: Matthies, M., Malchov, H. and Kriz, J. (eds) *Integrative Systems Approaches to Natural and Social Sciences – System Science 2000.* Springer-Verlag, Berlin, pp. 319–327.

2 Restricted Monopoly R&D Pricing: Uncertainty, Irreversibility and Non-market Effects

Robert D. Weaver[1] and Justus Wesseler[2]

[1]Department of Agricultural Economics, 207D Armsby Building, Pennsylvania State University, University Park, PA 16802, USA; [2]Environmental Economics and Natural Resources Group, Social Sciences Department, Wageningen University, Hollandseweg 1, 6706KN, Wageningen, The Netherlands

Introduction

The design of a property rights protecting system determines what kind and at what costs R&D outputs are generated. Further, this may differ between conventional and transgenic crops (Goeschl, 2005). When innovations in agriculture biotechnology offer enhanced returns to users such as transgenic crops, and where use can be restricted through intellectual property rights (IPRs) such as patents, a return to the innovation can be claimed and used to finance R&D. The feasibility of this approach to financing R&D depends on the IPR establishing monopolistic pricing power for the technology developer. Under perfect monopoly, the innovator will price the technology where marginal costs equal marginal revenue, and the innovation will be drastic. This type of model has been applied to the welfare analysis of agriculture biotechnology (see, for example, Falck-Zapeda *et al.*, 2000).

Most of the innovations in transgenic crops are non-drastic. The transgenic crops technology competes with cultivation techniques for non-transgenic crops, e.g. competitors in the pesticide industry have lowered prices for pesticides in response (Scatasta and Wesseler, 2004).

Weaver and Kim (2002) reconsidered pricing power under a monopoly that is restricted by the presence of alternative technologies, showing that this implies a threshold price that restricts the range of pricing under the monopoly grant of a patent confirming the non-drastic characteristic of the technology. This general result has been presented by Dasgupta and Stiglitz (1980). Weaver and Kim consider the specific case where the threshold price is defined by the characteristics of the alternative technology, as well as by their farm-specific implications. This implies that optimal pricing under restricted monopoly is not uniform, but instead would be a contract-based pricing scheme where price is non-linear in the incentives of the patented and the alternative technology.

From an *ex-ante* perspective, the attention has focused on the strategic investment in R&D under uncertainty and irreversibility. Different types of real option models have been employed to analyse the optimal investment strategy for R&D. The uncertainty about benefits and costs of R&D investment provides incentives for delay (see, for example, Trigeorgis, 1996a; Pennings and Lint, 1997; Weeds, 2000). These types of models do not explicitly consider the pricing power of a monopolistic technology provider facing heterogeneous technology adopters.

In this chapter, the Weaver and Kim (W&K) framework for pricing of innovation is extended to incorporate uncertainty and irreversible costs as well as benefits. We combine the insights of the deterministic model of Weaver and Kim by considering *ex-ante* uncertainty and irreversibility of the technology adopter and the implications for monopolistic pricing. First, we briefly review the model by Weaver and Kim and include in a second step uncertainty and irreversibility in a real options framework for considering the timing of investment in or adoption of the innovation. Based on this framework, we evaluate pricing and adoption implications of the evolution of incentives, irreversible costs and benefits, as well as of the returns to the innovation.

Framework for Transgenic Innovation Pricing

To summarize the W&K framework briefly, suppose the innovator is also the supplier of the seed. The private returns to a seed-based innovation follow directly from revenue from the sale of the seed. Seed supplier profits are defined:

$$\pi_s^i \equiv w_c^i \sum_{j=1}^{J} s_c^i - c_s^i \left(s_c^i \right) \tag{1}$$

where J is a pre-determined target market population of producers of the cth crop using the ith technology, using seed s_c^i sold at a uniform price w_c^i, produced by the supplier at cost $c_s^i \left(s_c^i \right)$. Weaver and Kim define the production function for cth crop output y_{jc}^i (quantity per land area, e.g. ha) from the jth farm operating the ith technology that is conditioned by farm-specific quasi-fixed and fixed input flows represented by a vector, θ_j, by a stochastic shock, ε_j, generated by a density function $g(\varepsilon_j | 0, 1)$, and by a vector of inputs, x_c^i. For our purposes, we do not consider further the density function $g(\varepsilon_j | 0, 1)$, as we later introduce uncertainty in the form of a stochastic process. The input vector includes inputs relevant for the ith technology, though some elements may also be relevant for other technologies. Specific technologies are differentiated by unique attributes such as planting flexibility and management intensity. The crop output per land area production function is defined as $y_{jc}^i = y_{cj}^i(x_{cj}^i, \theta_j)$ and producer profit per land area for crop c produced with technology i as $\pi_{jc}^i \equiv p_{jc}^i y_{jc}^i - r_{jc}^i x_{jc}^i - w_c^i \delta_{jc}^i$, where p_{jc}^i is the output price that is allowed to be technology and farm

differentiated, r_{jc}^i is the input price vector, δ_{jc}^i is the seeding rate per land area, and w_c^i is the price paid for ith seed type for crop c. Here, we limit our consideration to a uniform seed price, w_c^i. Sequential planting decisions are specified where selection of the crop area and production plans are made conditional on earlier stage selection of optimal technology for each crop. Choice of technology results in a value function indicating the maximum value per land area unit for each crop c:

$$V_{jc}^i(\rho_{jc}^i) = V_{jc}^i(p_{jc}^i, r_{jc}^i, w_c^i, \delta_{jc}^i, \theta_j) \equiv \max \mathrm{EU}(\pi_{jc}^i)$$

$$\text{where } \pi_{jc}^i \equiv p_{jc}^i y_{jc}^i - r_{jc}^i x_{jc}^i - w_c^i \delta_{jc}^i$$

$$\text{s.t. } y_{jc}^i = y_{jc}^i(x_{jc}^i, \theta_j) \tag{2}$$

where a farm-specific vector of determinants of value is defined as an 'incentives' vector $\rho_{jc}^i \equiv [p_{jc}^i, r_{jc}^i, w_c^i, \delta_{jc}^i, \theta_j]$. The set (I) of all economically feasible alternative technologies for crop c is defined as those technologies i' for which $V_{jc}^{i'} > 0$ at the prevailing $\rho_{jc}^{i'}$. The producer's relative net benefit for technology i versus technology i' for the same crop is $\omega_{jc}^i = V_{jc}^i - V_{jc}^{i'}$. *Local dominance* of technology i for farm j follows if:

$$\omega_{jc}^i \equiv V_{jc}^i - V_{jc}^{i'} > 0 \quad \forall i' \neq i \in I \tag{3}$$

Technology i is globally dominant across the set of J farmers if Equation 3 holds for all $j \in J$. Following this notation, the *relative willingness to pay* for technology i for crop c versus the second best alternative is defined as:

$$\omega_{jc}^i \equiv V_{jc}^i - V_{jc}^{i'} > 0 \quad \forall [i, c] \neq [i', c'], \ c \in C, \ i \in I \tag{4}$$

where the value functions are continuous in their incentive vectors $\rho_{jc}^i = (p_{jc}^i, r_{jc}^i, w_c^i, \delta_{jc}^i, \theta_j)$ and $\rho_{jc}^{i'} = (p_{jc}^{i'}, r_{jc}^{i'}, w_c^{i'}, \delta_{jc}^{i'}, \theta_j) \forall i'$. Weaver and Kim note that according to Equation 4, the adoption decision, the demand for the ith technology, is not continuous in the incentives $(\rho_{jc}^i, \rho_{jc}^{i'})$. Further, Equation 4 can be viewed as a measure of the monopoly pricing power and implies that power is restricted by the presence of alternative technologies and crops.[1]

Real Options Perspective on Adoption of Transgenic Innovations

To proceed, we extend this deterministic theory to incorporate the case where costs and benefits associated with the technology may be both irreversible and uncertain. Under these conditions, flexibility and waiting have value, as has been recognized in the real options literature (see, for example, Dixit and Pindyck, 1994; Trigeorgis, 1996b; Merton, 1998). Under uncertainty and irreversibility, the producer may find advantage in postponing adoption as information evolves and uncertainty resolves, reducing the implications of irreversible cost and increasing those of irreversible benefits. It follows that in this case, the investment decision is best considered within a real options framework. We explore this setting and consider its implications for adoption and pricing of innovations. We continue to consider the problem within the setting of transgenic crops and pricing of seeds as considered in the previous section.

Two salient features of transgenic seed involve irreversibility and uncertainty. First, transgenic seed innovations include an array of changes in production practices involving both changes in the efficient levels of variable inputs and changes in quasi-fixed inputs and possible irreversible external changes. Thus, implementation of the technology selection involves investments that can be assumed to be irreversible. The second salient feature follows directly, i.e. the investment or adoption decision results in changes in uncertain returns that span future time periods. As noted by Weaver and Kim, the uncertain stream of returns involves both private pecuniary and public types of returns. The uncertainty characterizing the decision goes beyond the stochastic nature of production. Given the multiple period return stream associated with adoption, the choice is made in a context of uncertain future input and output prices, as well as input availability. In sum, these sources of uncertainty affect the net benefit stream that can be expected from the irreversible investments that constitute the act of adoption.

With respect to irreversible external effects of the adoption decision, consider the case of herbicide-tolerant soybeans. In this case, changes in field practices, pesticide use and pest management practices can be expected (see Carpentier and Weaver 1997; Carpenter 2001). These changes follow from irreversible investments in know-how, management practices and related equipment. Importantly, the result is a change in irreversible benefits, R, and irreversible costs, I. Both benefits and costs may incorporate private, or internal, changes such as farm operator health, and public,

or external, components such as impacts on soil structure and ecology or biodiversity.

The external costs of adopting transgenic crops have been considered extensively in the literature. O'Shea and Ulph (2002) discuss the impact on insects which may translate into loss of bird life. They analyse policy approaches that would internalize these effects. Morel et al. (2003) use a real option model to include irreversible environmental effects. Wesseler (2003), using a similar model, includes irreversible environmental benefits. Demont et al. (2005a) use a combined real option and partial equilibrium model to analyse the welfare effects of introducing herbicide-tolerant sugarbeet in the European Union (EU) including external irreversible costs and benefits. Morel et al. (2003) and Wesseler (2003) assume competitive pricing of the technology, while Demont et al. (2004) assume a monopolistic technology supply. Demont et al. (2005) compare different types of irreversibility and their impact on the value of transgenic crops. None of these studies specifically address the pricing of a new technology under uncertainty and irreversibility as is considered here. In addition to the internal irreversible effects due to changes in pesticide use, internal irreversible costs may be necessary to incur due to changes required in the composition of quasi-fixed inputs. For example, herbicide-tolerant soybeans allow no-till (zero tillage) planting and a higher planting density, implying that different soybean planting machinery may be optimal. When such subsidiary investments are required, the relative willingness to pay ω_{jc}^i is reduced. More importantly, a longer-term view is required to support the adoption decision.

In the setting of transgenic crop adoption, managers face stochastic prices and input quantities. The harvest price for soybeans and fertilizers, pesticides, fuel and other farm inputs is not known with certainty at the date of adoption. Importantly, when a technology is not universally dominant, strategic response by vendors of an entire, or components of, competing technology can be expected to result in price reductions. The timing and magnitude of these competitive responses are uncertain for the farmer.

Given these conditions, the timing of the act of adoption becomes an important control variable for the farmer. Irreversible investment requirements, uncertain pay-offs and the option to adopt (flexibility) directly affect the willingness to pay for the new technology, implying that an extension of the deterministic framework in Equation 7 would be fruitful. In the literature on real option valuation, the opportunity to invest is valued in analogy to a call option in financial markets. The investor has the right, but not the obligation, to exercise his investment. This right, the option to invest (real option), has a value that results from the option owner's flexibility and is similar to the quasi-option value developed earlier by Arrow and Fisher (1974), Henry (1974) and Fisher (2000).

The agent is viewed as recognizing that information will change over time, and as having priors concerning the nature of those changes in the net present value and irreversible cost and benefit streams. The real options approach supposes that an investor has the opportunity to buy (go long) a call option, valued at F. This option provides the right, but not the obligation, to adopt transgenic crops. The value of this call option for adoption of a technology i by incurring irreversible costs, I, and benefits, R, for crop c by farmer j can be written in general form as:

$$F = F(\Omega, I, R, t) \geq 0 \qquad (5)$$

where the value of the option depends on: (i) the *relative discounted* willingness to pay for the new technology i for crop c defined as:[2]

$$\Omega_{j,t} = \int_{t=1}^{t=\infty} \omega_{jc}^i e^{\alpha t} e^{-\mu t} dt = \frac{\omega_{jc}^i}{\mu_{\omega_{jc}'} - \alpha_{\omega_{jc}'}}, \quad [3] \qquad (6)$$

where $\alpha_{\omega_{jc}^i}$ is the expected growth rate of the relative willingness to pay ω_{jc}^i and $\mu_{\omega_{jc}^i}$ interpretable as the risk-adjusted annual return derived from the capital asset pricing model (CAPM); (ii) the irreversible costs, I; and (iii) the irreversible benefits, R. As Equation 6 indicates, our model is autonomous, implying that the time variable can be dropped as a dependent variable.

Now, assume farmer j must choose between buying transgenic crops at the market or producing them on his farm. We want to derive the value of the option to 'adopt transgenic technology' or, equivalently, to produce the transgenic crop on the farm. To evaluate the investment decision, the investor is viewed as constructing a portfolio of a long in the call option that is balanced by sale of assets with known value. Specifically, the investor would like to balance this sale of assets such that the portfolio has zero risk. Suppose this can be done by selling forward (short): (i) m units of Ω, the *relative discounted* willingness to pay of the new

technology i for crop c; and (ii) n units of net irreversible costs, IR.[4] To proceed, the superscript i for technology and the subscripts c for crop and j for farmer are dropped hereafter to improve the readability. For the following, we will always assume that $F(\Omega, IR)$ is the value of the option to produce a crop, c, with specific technology, i, for a given farmer, j.

We next define the dynamic characteristics of Ω, and IR. We define $a_\Omega \equiv \alpha_{\omega jc}^i$ and assume Ω and IR follow uncorrelated geometric Brownian motion processes:

$$d\Omega = \alpha_\Omega \Omega + \sigma_\Omega \Omega dz, \quad \text{and}$$
$$dI = \alpha_{IR} IR + \sigma_{IR} IR dz \qquad (7)$$

Note, the geometric Brownian motion is a non-stationary, continuous time stochastic process in which α_i is a constant drift rate, σ_i is the constant variance rate, and dz is the Wiener process, with $E(dz) = 0$ and $E(dz)^2 = dt$. The geometric Brownian motion is the limit of a random walk (Cox and Miller, 1965), hence it is consistent with the assumption of log-normality of the stochastic variable with zero drift and is often chosen by economists because of its analytical tractability. The expected value of this process grows at the rate α. A constant positive growth rate assumes that the willingness to pay grows continuously over time and the innovation becomes increasingly attractive for adoption. This does not exclude the possibility that in any future period, $V_{jc}^{i'}(\rho_{jc}^{i'}) > V_{jc}^{i}(\rho_{jc}^{i})$. The implicit assumption made is that technology adopters can switch between the two technologies without bearing any additional costs.

At any point in time, the gross return of a portfolio composed of a call option to invest that is financed by shorts in Ω, and IR is simply $F - m\Omega - nIR$. The change in this gross return over a short time interval dt is $d(F - m\Omega - nIR)$. In order to finance the short positions in the portfolio, an equivalent long must be sold and a dividend paid. An investor willing to buy the long position and hold it over a short time interval will require a rate of return equal to the risk-adjusted rate of return of each short sale. This risk-adjusted rate of return equals the growth rate α_i, $i = \Omega$, IR, plus any dividend stream, δ_i,[5] paid to the buyer of the long. Note that by selling the shorts, the long position is covered, i.e. the risk of the bag is eliminated. From this perspective, the portfolio is viewed as a riskless portfolio.

To define the net return from holding the riskless portfolio over a short time interval:

$$dX \equiv d(F - m\Omega - nIR) - (m\delta_\Omega\Omega - n\delta_{IR}IR)dt \quad (8)$$

The change in value of the gross return of the portfolio over an infinitesimal small time step, dt, using Itô's lemma is:

$$\begin{aligned}
d(F - m\Omega - nIR) = &(\partial F/\partial\Omega - m)d\Omega \\
&+ (\partial F/\partial IR - n)dIR \\
&+ 1/2(\partial^2 F/\partial\Omega^2\sigma^2_\Omega\Omega^2 \\
&+ \partial F^2/\partial^2_{IR}\sigma^2_{IR}IR^2)dt
\end{aligned} \quad (9)$$

Note that the first two total differentials on the right-hand side are stochastic, reflecting the stochastic properties of the differentials in willingness to pay and net irreversible costs. However, by choosing m and n, it is clear that the stochastic effects on the portfolio return can be eliminated. We assume the investor has an interest in optimizing the gross return, for which the necessary conditions require m, and n are set equal to $dF/d\Omega$ and dF/dIR, respectively.

We are now in a position to reconsider the investment decision. The choice to invest can be viewed as a choice to exercise the option or, equivalently, sell the riskless portfolio. From the investor perspective, this will be done only when the net return is less than the market rate of return for riskless assets. In equilibrium, arbitrage will drive the rate of return of this riskless portfolio to equal the market rate of return, r, for a riskless asset, $r(F - m\Omega - nIR)$. The arbitrage equilibrium can be written using Equations 8 and 9 as:

$$\begin{aligned}
&1/2(\partial^2 F/\partial\Omega^2\sigma^2_\Omega\Omega^2 + \partial F^2/\partial^2_{IR}\sigma^2_{IR}IR^2)dt \\
&- m\delta_\Omega\Omega - n\delta_{IR}IR = r(F - m\Omega - nIR)
\end{aligned} \quad (10)$$

or by collecting terms

$$\begin{aligned}
&1/2((\partial^2 F/\partial\Omega^2)\sigma^2_\Omega\Omega^2 + (\partial F^2/\partial^2_{IR})\sigma^2_{IR}IR^2)dt \\
&+ (r - \delta_\Omega)(\partial F/\partial\Omega)\Omega \\
&+ (r - \delta_{IR})(\partial F/\partial IR)IR - rF = 0
\end{aligned} \quad (11)$$

The partial differential Equation 11 includes the two independent variables Ω and IR. The $(F, \Omega - IR)$ space can be divided into two regions: one region where it is optimal not to adopt the new technology (not exercise the option) and one

where it is optimal to adopt the new technology immediately (exercise the option). The boundary between the two regions is defined as a set of points, where the value of the option to invest, F, and the value of the immediate investment or adoption, $\Omega - IR$, are equal. As long as the value of the option to invest is greater than the value of immediate investment, it pays to wait. Hence, one of the conditions that define this boundary between the two regions of the $(F, \Omega - IR)$ space is the value matching condition:[6]

$$F(\Omega, IR) = \Omega - IR \quad (12)$$

That is, the value of the option is equal to the net value of immediate investment. The boundary between the two regions is not determined by Equation 12. There are several possible boundaries that satisfy Equation 12. The problem of finding the correct one, (the 'free boundary' problem) can be solved by adding two conditions. At the boundary, the value of the option has to be continuous and smooth to assure uniqueness (Dixit, 1993). The following conditions ('smooth-pasting') ensure this continuity and smoothness:

$$\partial F/\partial\Omega = 1 \quad (13)$$

$$\partial F/\partial IR = -1 \quad (14)$$

Finding a solution for F in Equation 11 that satisfies the conditions of Equations 12–14 is complicated by its elliptical form. This problem can be avoided by further simplifying the option value function. If we assume the option value function is homogenous of degree 1 in Ω and IR, following McDonald and Siegel (1986), we get the following boundary property results:

$$\left(\frac{\Omega}{IR}\right)^* = \frac{\beta_1}{\beta_1 - 1} \quad (15)$$

where

$$\begin{aligned}
\beta_1 = &\frac{1}{2} - \frac{\delta_{IR} - \delta_\Omega}{\sigma^2_\Omega + \sigma^2_{IR}} \\
&+ \sqrt{\left(\frac{\delta_{IR} - \delta_\Omega}{\sigma^2_\Omega + \sigma^2_{IR}} - \frac{1}{2}\right)^2 + \frac{2\delta_{IR}}{\sigma^2_\Omega + \sigma^2_{IR}}}
\end{aligned}$$

is the positive root of the solution to Equation 15.

As Ω is a constant multiple of ω (see Equation 6), ω follows a geometric Brownian motion with the same drift parameter α_Ω and variance parameter σ_Ω.[7] Using Equation 6, Equation 15 can also be written as:

$$\left(\frac{\omega}{IR}\right)^* = \frac{\beta_1}{\beta_1 - 1}(\mu_\Omega - \alpha_\Omega) \qquad (16)$$

where we define $\mu_\Omega \equiv \mu_{\omega_{jc}^i}$ for convenience. Note, the expression in Equation 15 is interpreted as the hurdle rate and in Equation 16 as the annualized hurdle rate. It defines a threshold or hurdle for Ω/IR that determines the decision of immediate adoption.

Interpretation of the adoption rule

If we compare the results of Equation 15 with Equation 4, we can immediately observe that under uncertainty, irreversibility and flexibility, the willingness to pay for the new technology will be smaller than in the deterministic setting. Under the deterministic pricing system, a positive value for ω is necessary if a technology fee or premium over the price of alternative technologies is to be feasible. Under uncertainty, irreversibility and flexibility, ω must be greater than ω^*, if the technology is to be adopted for exogenously given IR. Note, in the deterministic case, $\omega > 0$ implies adoption. Under the real option approach, we also consider irreversible benefits and costs and the rule changes. In effect, Equation 16 is the new decision rule. Equation 16 defines the boundary relationship between ω and IR that divides the (ω, IR) space into adoption and non-adoption regions. Suppose a farmer has exogenously given IR, then Equation 16 defines a ω^* on the boundary. If $\omega > \omega^*$, then the real option decision rule would signal the farmer to adopt the technology immediately. Thus, as IR increases, ω^* must increase according to Equation 16. The effect is that the adoption signal may change from adopt immediately to postpone adoption.

Implications for the pricing power

In the deterministic case, if $\omega(w) > 0$, pricing power exists. The seed price w can increase while adoption maintains until $\omega(w) = 0$. If net irreversible costs and uncertainty are considered,

$$\left(\frac{\omega}{IR}\right)^* \equiv \gamma \qquad (17)$$

and by Equation 20

$$\gamma = \frac{\beta_1}{\beta_1 - 1}(\mu_\Omega - \alpha_\Omega) \qquad (18)$$

for any exogenously given IR,

$$\omega^* \equiv \gamma \overline{IR} \qquad (19)$$

pricing power exists if:

$$\omega(\underline{w}) > \omega^* = \gamma \overline{IR} \qquad (20)$$

and, again, technology vendors can increase the seed price until $\omega(\underline{w}) = \omega^* = \gamma \overline{IR}$.

A second important result is the effect of irreversible benefits. If there are no irreversible costs, yet irreversible benefits exist, then current period willingness to pay would be $\omega + R$, implying that the pricing power of the technology supplier would increase. If there are irreversible costs, the impact of one unit of irreversible benefits is the ratio $\beta_1 / (\beta_1 - 1) > 1$. Clearly, this case clarifies the interest of the technology vendor in promoting the magnitude of R, while downplaying the existence of I. This strategy clearly increases the pricing power. Again, an important example is the case of soybeans, where adoption of zero tillage systems increased with the introduction of herbicide-tolerant soybeans providing environmental benefits of reduced soil erosion, pesticide and fuel use (Fawcett and Towery, 2003). An adoption of zero tillage requires a change in land management practices including farm equipment (Lambert and Lowenberg-DeBoer, 2003).

Quantitative Relevance of the Irreversibility Effect

In the previous section, we have shown that irreversible costs and benefits will have an effect on the pricing power of the technology supplier. In this section, we will apply the theoretical model to the case of herbicide-tolerant soybeans in Pennsylvania. Recall that the ratio $(\omega/IR)^* \equiv \gamma$ is interpretable as the hurdle rate. An increase in the drift for the willingness to pay leads to an increase in the hurdle rate (see Equation 16). This highlights an important trade-off for the pricing policy of the new technology. An increase in future benefits of the technology increases the incentive to wait, reducing earlier adoption. Importantly, the hurdle

Table 2.1. Parameter ranges used for Monte Carlo simulation.

α_{IR}	α_{Ω}	σ_{IR}	σ_{Ω}	μ
−0.05 to 0.08	−0.05 to 0.08	0.10 to 0.60	0.10 to 0.60	0.05 to 0.15

rate increases sharply at growth rates above 5%. Two effects can explain the increase in the hurdle rate with an increase in the drift rate. Higher drift rates increase the value of future benefits from the technology. The value of the option to adopt herbicide-tolerant soybeans increases. The increase in the drift rates also increases the value of immediate adoption as the opportunity costs decrease due to a decrease of the discounting effect (see Equation 9). The former effect is greater and results in an increase in the hurdle rate. On the other hand, an increase in the drift rate of the net irreversible costs reduces the hurdle rate. A future increase in the costs reduces the value of waiting. Both drift rates can be positive or negative. The drift range of the willingness to pay is most likely to be positive. Farmers improve the efficiency of planting herbicide-tolerant soybeans from learning-by-doing. A decrease in herbicide prices as observed in the USA will improve the competitiveness of the alternative technology, but at the same time prices for glyphosate herbicides decreased, reducing this effect.

As mentioned earlier, the suppliers of alternative technologies have reacted to the introduction of herbicide-tolerant soybeans by reducing the prices related to the alternative technology, e.g. for herbicides. There may also exist the case where the alternative technology pricing strategies result in $\omega_{jc}^i < 0$, implying that the alternative technology becomes dominant over the new technology. By design, the model assumes that the adopter may switch between technologies without bearing any additional costs. The overall impact of this specification is that *ex-ante*, as α_{Ω} is smaller or even negative, the hurdle rate is reduced, early adoption is increased, immediate pricing power is increased, though future pricing power is reduced.

The assumption of a constant drift rate implies a continuous increase in the willingness to pay over time. Another possible scenario is a parallel time path of the two incentive vectors. In this case, the expected value of a change in the willingness to pay is zero ($\alpha_{\Omega} = 0$), and only uncertainty remains and the pricing power of the technology provider remains constant.

Recall that the variance rate in Equation 7 is interpretable as a measure of uncertainty. An increase in the variance rate increases the hurdle rate. Greater uncertainty about future net irreversible costs and/or willingness to pay increases the option value. Alternatively, an increase of the risk-adjusted rate of return reduces the hurdle rate, as the value of the future decreases. As the risk-adjusted rate of return approaches the drift rate of the willingness to pay from the right, the hurdle rate increases to infinity and it always pays to wait as the value of the future becomes indefinitely large. The smaller the difference between the drift rate of the willingness to pay and the risk-adjusted rate of return, the lower is the convenience yield, δ. It follows that the lower the convenience yield, the lower is the pricing power of the technology provider.

The actual pricing power of the monopolistic technology provider depends on the parameter values defining β. The effects of parameter changes, such as an increase in the drift rate or uncertainty about future benefits, on pricing power are analysed by using Monte Carlo simulation for selected parameter ranges as reported by Weaver and Wesseler (2004) in more detail. Here we present the results of the simulation. The ranges for parameter values are shown in Table 2.1. For all parameters, a uniform distribution was assumed.

Parameter combinations violating model assumptions were deleted from the sample. A linear function approximating the hyper-plane was estimated. The results of the linear approximation are provided in Table 2.2. As the goodness of fit with 97.8% is already very high, we do not report results including second order effects. The negative coefficients on the two drift rates indicate the positive impact of an increase in future annual benefits or net irreversible costs on pricing power.

Calculating the drift rate elasticity of the pricing power at the mean, reported in Table 2.3, provides a value of about −0.039. A 1% increase in one of the drift rates reduces the pricing barrier γ by about 0.039%, but as this also increases the present

Table 2.2. Linear approximation of the pricing power including first order effects.

Dependent variable: γ

Method: least squares

Sample: 137,374

Included observations: 37,374

$$\gamma = C(1) + C(2)^*\alpha_{IR} + C(3)^*\alpha_\Omega + C(4)^*\sigma_{IR} + C(5)^*\sigma_\Omega + C(6)^*\mu$$

	Coefficient	SE	t-statistic	Probability
C(1)	−0.112646	0.000384	−293.0278	0.0000
C(2)	−0.759945	0.002033	−373.8811	0.0000
C(3)	−0.757656	0.002031	−373.1018	0.0000
C(4)	0.387532	0.000517	749.2061	0.0000
C(5)	0.386833	0.000517	747.7261	0.0000
C(6)	1.515313	0.002704	560.4844	0.0000
R^2	0.978110	Mean dependent variable	0.294230	
Adjusted R^2	0.978107	SD dependent variable	0.097459	
SE of regression	0.014420			
Sum squared residual	7.770318			
Log likelihood	105404.8			

Table 2.3. Elasticities of pricing power at the mean level using first order results.

	α_Ω	α_{IR}	σ_Ω	σ_{IR}	μ
Elasticities	−0.0386	−0.0387	0.4602	0.4610	0.5150

value of the project and this even more strongly (see Equation 6), it decreases the pricing power of the monopolist $(\partial(\beta/\beta - 1)/\partial\alpha > 0)$. An increase in future uncertainty increases the pricing barrier γ and reduces the pricing power of the monopolist. The variance rate elasticity of the pricing power at the mean provides a value of about 0.46. An increase in future uncertainty increases the pricing barrier γ but also reduces future benefits and increases the pricing power of the monopolist. The discount rate elasticity of the pricing power at the mean provides a value of about 0.51. A change in the discount rate or the variance rate has an about ten times stronger impact on pricing power than a change in the drift rate. This provides stronger incentives for reducing future uncertainty related to a new technology than to increasing the speed of technology development.

Further empirical application of the model is limited by available data. However, net irreversible costs for herbicide-tolerant soybeans depend largely on the difference between the costs of a planter or a drill for a no-tillage and a tillage cropping system. This difference is in the range of about US$20,000 for a drill. The average annual costs are about US$3500. The irreversible costs per hectare depend on the hectares planted with herbicide-tolerant soybeans. In Pennsylvania, farmers planted on average about 70 acres with herbicide-tolerant soybeans in the year 1999. The average irreversible costs are about US$50 per acre in this case. If the hurdle rate is about 2, the increase in the gross margin per acre has to be about US$100. The stylized example presented here indicates that the pricing power depends on the area planted to soybeans and hence on farm size, as confirmed by Weaver (2005).

An increase in farm size increases the pricing power of the technology provider. The irreversibility effect increases the relevance of the farm size with respect to the pricing power. If the technology provides irreversible benefits, the net irreversibility effect has to be considered. As one unit of irreversible costs requires more than one unit of willingness to pay to compensate, it is obvious that one unit of irreversible benefits increases the willingness to pay by more than one unit.

Conclusions

This chapter finds that irreversibility and uncertainty of a new technology reduce the willingness to pay of the technology adopter and the pricing power of the monopolistic technology provider. The pricing power decreases with an increase in future benefits and future uncertainty, while the effect of a 1% change in uncertainty has a ten times stronger effect on the hurdle rate than the same change on the increase in future benefits.

The irreversible costs of the technology in the case of herbicide-tolerant soybeans depend on the farm size. Larger farms are found, *ceteris paribus*, to have a higher willingness to pay than smaller farms. The irreversibility effect is found to aggravate the effect of farm size on pricing power. In this case, the pricing power of the technology provider is reduced further and has a negative impact on incentives for R&D in this area. On the other hand, technology providers have the chance to increase their returns by horizontal and vertical integration with providers of the supplementary investment.

Many technologies provide irreversible benefits, e.g. reduced pesticide and fuel use. If the adopter of the technology also considers those social benefits in his private decision making, his willingness to pay for the technology increases. The importance of irreversible benefits is greater in the case where irreversible costs are present. This provides an economic argument that explains why technology providers emphasize the irreversible environmental benefits of their technology.

Notes

[1] Lapan and Moschini (2000) considered the case where a competitively supplied input demand is affected by adoption of an innovation showing that a monopolist's optimal pricing strategy would lead to incomplete adoption by a population of homogeneous producers.

[2] The relative willingness to pay includes changes in the value of the alternative technology induced by strategic responses of vendors of those technologies or their components, e.g. price changes for pesticides for non-transgenic crops. In general, recall that ω_{jc}^i is defined relative to the next best technology as in Equation 7.

[3] The motivation for choosing the risk-adjusted rate of return is that the risk of the additional benefits could be tracked with a dynamic portfolio of market assets. $\mu_{\omega_{jc}^i} = r + \phi\sigma\rho(\omega_{jc}^i, m)$, where r is the risk-free rate of return, ϕ the market price per unit of risk, σ the variance parameter and $\rho(\omega_{jc}^i, m)$ the coefficient of correlation between the asset or portfolio of assets that track ω_{jc}^i and the whole market portfolio. See Dixit and Pindyck (1994, pp. 147–150) for an elaboration of this assumption.

[4] We assume net irreversible costs IR as this will significantly simplify our analysis. A distinction between irreversible costs and irreversible benefits can be made, but an analytical solution to such a problem would be difficult to find.

[5] δ_i is the so-called convenience yield, that results from owning an asset instead of holding the monetary capital. The main reason for the convenience yield is higher flexibility in the case of machinery.

[6] Chapter 12 of Neftci (2000) provides a good introduction into boundary conditions for solving partial differential equations.

[7] Equation 6 states $\Omega_j = \omega_j/(\mu - \alpha)$, hence, as μ and α are constants,

$$d\Omega_j = d\left[\frac{\omega}{(\mu - \alpha_\Omega)}\right] = \frac{1}{\mu - \alpha_\Omega}d\omega$$

$$= \alpha_\Omega\frac{\omega}{(\mu - \alpha_\Omega)}dt + \sigma_\Omega\frac{\omega}{(\mu - \alpha_\Omega)}dz$$

and $d\omega = \alpha_\Omega\omega dt + \sigma_\Omega\omega dz$.

References

Arrow, K. and Fisher, A. (1974) Environmental preservation, uncertainty, and irreversibility. *Quarterly Journal of Economics* 88, 312–319.

Carpenter, J.E. (2001) *Case Studies in Benefits and Risks of Agricultural Biotechnology: Roundup Ready Soybeans and Bt Field Corn*. National Center for Food and Agricultural Policy, Washington, DC.

Carpentier, A. and Weaver, R.D. (1997) Damage control productivity: why econometrics matters.

American Journal of Agricultural Economics 78, 47–61.

Cox, D.R. and Miller, H.D. (1965) *The Theory of Stochastic Processes*. Chapman and Hall, London.

Dasgupta, P. and Stiglitz, J. (1980) Uncertainty, industrial structure and the speed of R&D. *Bell Journal of Economics* 11, 1–28.

Demont, M., Wesseler, J. and Tollens, E. (2004) Biodiversity versus transgenic sugar beets – the one Euro question. *European Review of Agricultural Economics* 31, 1–18.

Demont, M., Wesseler, J. and Tollens, E. (2005) Reversible and irreversible costs and benefits of transgenic crops. In: Wesseler, J. (ed.) *Environmental Costs and Benefits of Transgenic Crops*. Wageningen UR Frontis Series Vol. 7. Kluwer Academic Publishers, Dordrecht, The Netherlands, pp. 113–122.

Dixit, A. (1993) *The Art of Smoothpasting*. Harwood Academic Publishers, Chur, Switzerland.

Dixit, A. and Pindyck, R.S. (1994) *Investment under Uncertainty*. Princeton University Press, Princeton, New Jersey.

Falck-Zepeda, J.B., Traxler, G. and Nelson, R.G. (2000) Surplus distribution from the introduction of a biotechnology innovation. *American Journal of Agricultural Economics* 82, 360–369.

Fawcett, R. and Towery, D. (2003) *Conservation Tillage and Plant Biotechnology*. Conservation Technology Information Center, West Lafayette, Indiana.

Fisher, A. (2000) Investment under uncertainty and option value in environmental economics. *Resource and Energy Economics* 22, 197–204.

Goeschl, T. (2005) Do patent-style intellectual property rights on transgenic crops harm the environment? In: Wesseler, J. (ed.) *Environmental Costs and Benefits of Transgenic Crops*. Wageningen UR Frontis Series Vol. 7. Kluwer Academic Publishers, Dordrecht, The Netherlands, pp. 203–218.

Henry, C. (1974) Investment decision under uncertainty: the irreversibility effect. *American Economic Review* 64, 1006–1012.

Lambert, D. and Lowenberg-DeBoer, J. (2003) Economic analysis of row spacing for corn and soybean. *Agronomy Journal* 95, 564–573.

Lapan, H. and Moschini, G. (2000) Incomplete adoption of a superior innovation. *Economica* 67, 525–542.

McDonald, R. and Siegel, D. (1986) The value of waiting to invest. *Quarterly Journal of Economics* 101, 707–728.

Merton, R.C. (1998) Application of option pricing theory: twenty-five years later. *American Economic Review* 88, 323–349.

Morel, B., Farrow, S., Wu, F. and Casman, E. (2003) Pesticide resistance, the precautionary principle, and the regulation of Bt corn: real option and rational option approaches to decisionmaking. In: Laxminarayan, R. (ed.) *Battling Resistance to Antibiotics. An Economic Approach*. Resources for the Future, Washington, DC, pp. 184–213.

Neftci, S. (2000) *An Introduction to the Mathematics of Financial Derivatives*. Academic Press, New York.

O'Shea, L. and Ulph, A. (2002) Providing the correct incentives for genetic modification. In: Swanson, T. (ed.) *The Economics of Managing Biotechnologies*. Kluwer Academic Publishers, London, pp. 129–143.

Pennings, E. and Lint, O. (1997) The option value of advanced R&D. *European Journal of Operational Research* 103, 83–94.

Scatasta, S. and Wesseler, J. (2004) *A Critical Assessment of Methods for Analysis of Environmental and Economic Cost and Benefits of Genetically Modified Crops in a Survey of Existing Literature*. Presented at the 8th International Conference of ICABR, Ravello, 2004.

Trigeorgis, L. (1996a) The nature of option interactions and the valuation of investments with multiple real options. *Journal of Financial and Quantitative Analysis* 28, 1–20.

Trigeorgis, L. (1996b) *Real Options*. The MIT Press, Cambridge, Massachusetts.

Weaver, R.D. (2005) *Ex post* evidence on adoption of transgenic crops: US soybeans. In: Wesseler, J. (ed.) *Environmental Costs and Benefits of Transgenic Crops*. Wageningen UR Frontis Series Vol. 7. Kluwer Academic Publishers, Dordrecht, The Netherlands, pp. 203–218.

Weaver, R.D. and Kim, T. (2002) *Incentives for R&D to Develop GMO Seeds: Restricted Monopoly, Nonmarket Effects, and Regulation*. Presented at the 6th International Conference of ICABR, Ravello, 2002.

Weaver, R.D. and Wesseler, J. (2004) Monopolistic pricing power for transgenic crops when technology adopters face irreversible benefits and costs. *Applied Economics Letters* 15, 969–973.

Weeds, H. (2000) *Strategic Delay in a Real options Model of R&D Competition*. Warwick Economic Research Papers 576. Department of Economics, University of Warwick, UK.

Wesseler, J. (2003) Resistance economics of transgenic crops. a real option approach. In: Laxminarayan, R. (ed.) *Battling Resistance to Antibiotics. An Economic Approach*. Resources for the Future, Washington, DC, pp. 214–237.

3 Biotechnology and the Emergence of Club Behaviour in Agricultural Trade

Monika Tothova and James F. Oehmke

Department of Agricultural Economics, Michigan State University, East Lansing, MI 48824, USA

Perhaps the most contentious issue of the day stirring the agricultural trade – recently advancing to the World Trade Organization's (WTO) dispute settlement procedure – is the treatment of transgenic crops and genetically modified (GM) foods. The disagreement mainly concerns consumer acceptance of GM foods and the different approaches to government regulation of the agricultural biotechnology industry. The argument, earlier contained to a transatlantic clash between the European Union (EU)[1] and the USA, recently has spread to the rest of the world, including developing countries, some still challenged by ensuring food security for their populations (Bernauer, 2003; Tothova and Oehmke, 2004). This issue was raised most visibly in 2003, when Zambia rejected food aid in the form of GM maize even in the midst of crop failure and widespread hunger and malnutrition (Malawi and Zimbabwe, facing similar circumstances, originally rejected the aid but eventually accepted the maize if it was milled immediately to prevent GM seeds from entering the food system). In contrast, India and China appear to be forging ahead with GM food production and consumption, despite particularly notable opposition in India from non-governmental organizations (NGOs). The difference between southern Africa and India/China exemplifies the pattern of bifurcation noticed throughout the world. Countries are largely aligning themselves with one of two poles – they either align themselves with the EU in rejecting GM foods (i.e. applying the precautionary principle),

or they align themselves with North America in adopting GM food production and consumption (i.e. endorsing substantial equivalence) (Bernauer, 2003; Tothova and Oehmke, 2004). This chapter investigates the joint emergence of GM food standards and *de facto* 'clubs' of countries that adopt similar standards (or lack of standards) for GM crop production and consumption.

The foundation for disparity between the EU and the USA as principals in the 'disagreement' rests on the differences in respective consumer preferences (i.e. Perdikis, 2000; Lusk *et al.*, 2003; Moon and Balasubramanian, 2003, etc.), and results in divergent underlying regulatory approaches adopted by each country.

The USA, operating on the premise of scientific rationality (Isaac and Kerr, 2003), argues that current transgenic crops are 'substantially equivalent' to their non-transgenic counterparts and thus fall under the GATT definition of 'like products' – products with the same end use and identical tariff classification. The 'substantial equivalence' position means that no specific actions – either governmental or private – are necessary to determine acceptable tolerance level of GM foods in the system. Moreover, this position implies that genetic modification is a production process (rather than a product characteristic), and thus any 'discrimination' against GM foods is a trade barrier per WTO rules.

The EU, adopting the premise of social rationality and refusing substantial equivalence (Kerr, 2003; Isaac and Kerr, 2003), considers GM foods

---

Table 3.1. GMO regulatory approaches.

	The risk analysis framework (RAF)	
	Scientific rationality	Social rationality
General regulatory issues		
Belief	Technological progress	Technological precaution
Type of risk	Recognized Hypothetical	Recognized Hypothetical and speculative
Substantial equivalence	Accepts substantial equivalence	Rejects substantial equivalence
Science or other in risk assessment	Safety Health	Safety Health Quality 'Other legitimate factors'
Burden of proof	Traditional: innocent until proven guilty	Guilty until proven innocent
Risk tolerance	Minimum risk	Zero risk
Science or other in risk management	Safety or hazard basis: risk management is for risk reduction and prevention only	Broader socio-economic concerns: risk management is for social responsiveness
Specific regulatory issues		
Precautionary principle	Scientific interpretation	Social interpretation
Focus	Product-based, novel applications	Process- or technology-based
Structure	Vertical, existing structures	Horizontal, new structures
Participation	Narrow, technical experts Judicial decision making	Wide, 'social dimensions' Consensual decision making
Mandatory labelling strategy	Safety- or hazard-based	Consumers' right to know-based

Source: Isaac and Kerr (2003).

to be different from their conventional counterparts based on the process, opposes introduction of transgenic crops and GM foods into the food system, and has raised a number of legal and market barriers that have deterred trade.[2] For example, after the introduction of transgenic maize to the USA, maize exports to the EU fell by 95%. Perhaps the most important 'barrier' is the EU labelling standard: any food containing more than 0.9% transgenic material has to be clearly labelled 'This product is produced from GMOs'. Since both domestic and foreign producers are expected to meet the same standard, this requirement is not considered a barrier to trade by WTO rules.

The GM debate essentially boils down to contrasting two different regulatory approaches, each backed up by different international treaties: the Cartagena Protocol on Biosafety (to which the USA is not a signatory) endorses the precautionary principle, while the WTO agreements give their approval to scientific evidence – and indirectly endorse substantial equivalence. A summary of the two different regulatory approaches is shown in Table 3.1. Wishing to abstain from making any normative statements regarding the superiority of either regulatory approach, we will consider the precautionary principle and substantial equivalence as examples of horizontal separation which makes each regulatory approach different, and not directly comparable in terms of performance characteristics. Accordingly, we focus on the consequences of two opposing regulatory approaches for the world trading system.

We tailor the model to suit the USA and EU: the driving assumption in the chapter is that US (or North American in general) and EU consumers have different preferences concerning GM foods. These different preferences give rise to different domestic (regional in the EU case) policies regarding GM foods. We embed these different preferences in a Krugman-like monopolistic competition trade model (i.e. Dixit and Stiglitz, 1977; Krugman, 1979, 1980). Countries trade differentiated products, for example, food, differentiated by variety, e.g. apples and tomatoes. By assuming constant elasticity, we emphasize consumers' desire for variety (i.e. the more different kinds of food are available, the happier the consumer, *ceteris paribus*) in lieu of traditional economies of scale in production (since at the aggregate level it is unclear if there are any economies of scale in agriculture). We then examine the conditions under which countries will maintain different standards rather than harmonizing or reaching a compromise standard. When countries do not harmonize or compromise, the model gives rise to *de facto* 'clubs' of countries. Members of each club that have similar standards (or taste for similar regulatory approaches) engage in food trade with other members of that club. Differences in standards prevent members of different clubs from trading food products, so that there is no cross-club trade in food products.

The Model[3]

Products are differentiated not only in the variety space, but also in the standards[4] space. Every time a new variety is introduced, the quantity produced of each variety declines and the cost of production rises, but the consumers gain from greater variety, *ceteris paribus*. Once trade opens, each country produces a set of varieties, and wants to import other varieties that meet similar standards.

In autarky, each country would determine its own standard (or lack thereof) for defining GM and GM-free in order to meet domestic consumer preferences. We make several simplifying assumptions. Instead of modelling GM standards on a continuum of tolerance levels, we assume 'discreteness' of the standard: each variety can be produced with one of the two attributes – GM or GM-free.[5] We assume that the EU threshold effectively serves as a prohibition, since some countries do not allow import of GM maize and other crops (Perdikis, 2000). While a valid argument is that not

all food in the USA is genetically modified – and is not likely to be – we choose to generalize and model the US consumer (especially in autarky) as pro-technology oriented with preference for transgenic crops and GM foods.[6] We also hypothesize that both GM and GM-free varieties are priced equally. The proponents of GM argue that introduction of GM varieties is likely to result in decreased costs of production – and consequently lower prices. However, while GM varieties usually tend to be less labour intensive with a lower pesticide use, the cost of their seeds – incorporating the cost of research and development, as well as marketing of new varieties – is higher (e.g. Bernauer, 2003). As the share of cost of processed foods attributed to raw commodities decreases, prices of processed foods have not changed. In addition, current marketing channels do not allow separate channels to preserve the identity of GM and GM-free crops and, thus, if a country produces some GM crops, its entire production is considered 'contaminated'.

The model presented is static and instantaneous adjustments are made as needed. All agents have perfect foresight and perfect information. Issues related to asymmetric and hidden information are assumed away. Consumers have trust in labelling schemes (if present), and are able to distinguish between standards at no cost to them if standards for like products are in place. Transportation is costless. Firms in the model do not earn any excess profits, and government does not impose any taxes. Due to the zero profit assumption, producers are indifferent as to what standard is demanded from them, so that the standard is set at a level determined endogenously by the consumers' utility maximizing decision. The 'consumer decides' model seems to be well suited to describe the growing role of consumer concerns and demands in trade.

Each country, consisting of like individuals, has access to the same production technologies, and differs in its preference for transgenic crops and genetic modification (in a way that will be made more precise momentarily). We start with a basic Krugman model where goods are differentiated in the variety space only. Each consumer has the utility function:

$$U = \left(\sum_{j=1}^{N} (d_j)^\theta \right)^{\frac{1}{\theta}} \tag{1}$$

where

$$\theta = \left(1 - \frac{1}{\sigma}\right), \quad \sigma > 1 \qquad (2)$$

σ represents elasticity of substitution between varieties, assumed to be constant across pairs of varieties; d denotes consumption of the differentiated goods; and the subscript j implies the variety. The consumer derives utility from a large number of varieties (indexed 1 to $\mathcal{N}$). With scarce labour and positive fixed costs, it may be that only $n < \mathcal{N}$ varieties are actually produced. All varieties enter the utility function symmetrically. Each consumer is endowed with one unit of labour only; there is no capital so income consists entirely of wage earnings of w. The entire stock of labour is used in production; consumers do not derive any utility from leisure. The demand for the differentiated good is obtained through maximization of the CES utility subject to the budget constrained by ownership of one unit of labour and wage normalized to 1:

$$\text{Max} \left(\sum_{j=1}^{N} (d_j)^\theta \right)^{\frac{1}{\theta}}$$

$$\{d_j\}_{j=1}^{N} \quad \text{s.t.} \quad \sum_{j=1}^{N} p_j d_j = 1$$

Assuming an interior solution, the first-order conditions for the problem are:

$$\left(\sum_{j=1}^{N} (d_j)^\theta \right)^{\frac{1-\theta}{\theta}} (d_j)^{\theta-1} = \phi p_j \qquad (4)$$

$$\sum_{j=1}^{N} p_j d_j = 1 \qquad (5)$$

where ϕ is a Lagrange multiplier. Solving these two first-order conditions yields the demand for variety j where it is optimal to purchase all varieties available in equal quantities:

$$d_j = \frac{p_j^{-\sigma}}{\sum_{j=1}^{n} p_j^{1-\sigma}} \qquad (6)$$

σ represents the elasticity of substitution between pairs of varieties of the same product (i.e. food). The price elasticity of demand faced by producers is:

$$\sigma + \frac{p_j^{1-\sigma}}{\sum_{j=1}^{n} p_j^{1-\sigma}} (1-\sigma) \qquad (7)$$

As n (the number of actual varieties produced[7]) is large, we make the usual assumption that the firm disregards the second component in the elasticity term, and considers σ to be the elasticity of demand it faces (Helpman and Krugman, 1985).

Up to this point, the differentiated goods part of the model closely followed Krugman (1980). Now assume that each variety can have one of the two attributes (GM or GM-free). Consumers have preferences over the two attributes that imply the two are substitutes. In particular, we might view quantities of the variety in 'quality-adjusted units' (recall each country puts a different weight on different quality), so that

$$d_j^Q = d_j^F + \lambda^C d_j^G \qquad (8)$$

d_j^Q is the quantity of variety j that possesses quality Q, for $Q = F, G$ (F = GM-free, G = GM) and λ^C (C = EU or USA) is a country-specific 'discount' parameter that marks down the physical quantity if the variety consumed has an attribute different from the preferred one. We hypothesize that the EU has strong preference for GM-free varieties, and in the EU, $\lambda^{EU} < 1$. In the USA, we assume that $\lambda^{USA} > 1$: that in fact the consumer has a preference for GM varieties as they might represent advanced characteristics such as longer shelf life or improved taste (e.g. Flavr-Savr™ tomatoes, see Martineau, 2001). For example, due to stronger interest groups, it is also likely that US consumers may have a greater ability to internalize positive environmental benefits such as decreased pesticide use – or, put simply, might like new technologies. The European consumer, on the other hand, may be more likely to internalize the negative externality of endangered environment in the form of possible accidental release of the gene to the environment.

The linear lower tier subutility functions where differing qualities of any variety are substitutes, and increasing returns to scale in production ensure that only one quality of each variety will be consumed. To see this, think of consumers constructing their own consumption of variety j by purchasing the two different components and putting them together. Using this interpretation, the price of a unit of variety j is the minimum cost of creating the variety from the two component parts. If we let p_j^F and p_j^G represent the prices of the two components (although both are priced equally, we keep the superscripts for the sake of exposition), then the price of the bundle is $p_j = \min$ $(p_j^F, p_j^G/\lambda^C)$. This follows from the fact that it requires $1/\lambda^C$ units of the goods with quality of GM to provide the same utility as one unit of the goods with the quality of GM-free. If $p_j^F < p_j^G/\lambda^C$, then consumers only demand the quality of GM-free for variety j, where superscript C on the preference parameter λ indicates country. If the inequality is reversed, they demand only the quality of GM for variety j, and if the two are equal, consumers are indifferent. Since the process of the constituent components can, in principle, vary by variety, it is possible that some varieties will be constructed of only GM-free, others of only GM. As prices are the same, $\lambda^{EU} < 1$ and $\lambda^{USA} > 1$, GM-free prevails in the EU ($p_j^F < p_j^G/\lambda^C$), while GM succeeds in the USA, ($p_j^F > p_j^G/\lambda^C$).

Production follows a Krugman model of monopolistic competition, assuming fixed and marginal costs are identical across GM and GM-free varieties. Increasing returns in the differentiated sector are internal to firms – an initial outlay of labour ('fixed cost') is needed to start up production, resulting in decreasing average cost; marginal costs are constant within a (variety, attribute) combination. Firms are assumed to be symmetric, resulting in the same price and output across varieties. The presence of the scale economies ensures that in equilibrium only a finite number of varieties are produced, each firm produces a different variety and each variety will be produced to one standard (GM or GM-free) only. Each firm producing up to a certain specification faces the same cost function regardless of its location. All varieties are perfect substitutes in production. The number of varieties produced within the economy is determined by the number of firms.[8] Firms can enter freely into the industry. The usual profit maximization and zero profit entry conditions apply regardless of the chosen attribute which, along with consumer demands, determine price and production levels. The equilibrium in production is described by the number of firms and the price level. Since only labour is used in the production, following Krugman (1980), we specify the cost functions:

$$c_i(y_i) = wl_i(y_i) = w(B + y_i M) \qquad (9)$$

where $l_i(y_i)$ is the labour requirement to produce y_i units of variety i regardless of the attribute, B is the fixed labour component, and M is the marginal labour component for producing any variety.

Letting the wage equal 1, we set marginal revenue equal to marginal cost to find the profit-maximizing price charged for variety j:

$$p_j = \frac{\sigma}{\sigma - 1} M \qquad (10)$$

Since all varieties with a given attribute have the same marginal cost, the price of all varieties with the given attribute will be the same. By free entry, profits are driven to zero, so that average cost equals price. After solving the model, this implies that each firm will hire $F\sigma$ workers. If the amount of labour available is L, then the total number of varieties regardless of the attribute will be:

$$n = \frac{L}{B\sigma} \qquad (11)$$

In the basic Krugman model, marginal costs impact price, while fixed costs together with the size of the economy influence the number of varieties produced in an economy.

For a single (autarkic) economy with identical consumers, only the GM-free product is produced if $\lambda < 1$ (EU). Only the GM product is produced if $\lambda > 1$ (USA). Consequently, the indirect utility function in the EU ('GM-free' country) is:

$$V^F = \frac{1}{p} n^{\frac{1-\theta}{\theta}} \qquad (12)$$

In the USA ('GM' country), the indirect utility becomes:

$$V^G = \frac{\lambda}{p} n^{\frac{1-\theta}{\theta}} \qquad (13)$$

Superscripts serve merely notational purposes, F indicates that only GM-free is produced, G indicates that only GM is produced. Notice that λ (the weight put on the GM variety, i.e. the marginal rate of substitution between different attributes of the same variety) is listed without a country superscript due to a single-country case. Define $W(\lambda) = \max(V^F, V^G)$ to be social welfare as a function of λ. The graph $W(\lambda)$ is shown in Fig. 3.1. For $\lambda < 1$, the country maximizes welfare by producing only with the GM-free attribute. As λ exceeds 1, only varieties with the GM attribute are produced. Social welfare increases as a function of λ since the value to consumers of the GM attribute increases as λ increases.

'Opening trade' in a one-factor model coincides with Krugman (1980), and translates into a larger factor (labour) pool, as well as a larger

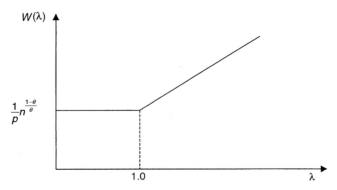

Fig. 3.1. Social welfare as a function of weight put on the consumption of the GM version: autarky.

product market to supply. Constant elasticity of demand ensures that the price levels (or outcomes of the profit maximizing conditions) do not change. The growth in the size of the market measured as an increase in the labour force does not influence the individual firm's output, but divides it among a larger number of consumers, resulting in lower per capita consumption of any variety. Also, due to the constant elasticity assumption, the entire increase in the stock of labour in the integrated economy is directed into production of varieties not existing in autarky, and does not increase the amount of labour directed to existing varieties. Intuitively, the basic Krugman model claims the trade is good as it (at least) increases the number of available varieties (the case in this model), and consumers in the integrated economy benefit from a larger number of varieties available to them at the price identical to the autarky price. Therefore, having more varieties available in an open trade than in autarky is a sufficient reason to trade and integrate (Krugman, 1980).[9]

In a free trade situation, a country that can import one variety (apples) from another country can reallocate the resources previously used in the production of apples to production of a new variety (kiwis, tangelos and strazzberries) and consumers would benefit from the introduction of a new variety. However, in order to import apples, the country must find an exporter whose GM/GM-free standards match (or exceed) the domestic standards. One way to accomplish this is by adapting the domestic standards to world market conditions/standards, but this entails a decrease in utility (from apple consumption at suboptimal standards) since the standards were originally set to meet domestic consumer preferences. Thus, the country must trade-off the utility gained from an increase in the

variety of food available with the loss in utility from suboptimal standards.

Two cases are theoretically possible. First, the EU and USA attempt to adopt a standard of one or the other country. While we recognize that unilateral harmonization is constrained by opposing consumer preferences and might be politically infeasible, we use it to demonstrate a general case that standardization does not always have to be welfare improving. Secondly, the EU and USA trade their respective goods without agreeing on a common standard, giving their consumers access to a larger than autarky number of varieties, albeit some at discounted utility. This non-harmonized trade scenario is always welfare-improving relative to autarky, but might not always be politically feasible due to domestic constraints and likely threats of trade disputes.[10]

For open trade scenarios, we assume that consumers in each country are able to distinguish between qualities, GM and GM-free goods are perfectly traceable, and their segregation is guaranteed. Labelling schemes, if present, are costless. Define n to be number of varieties produced in the EU, and n^* to be number of varieties produced in the USA. If trade is permitted, and all goods are standardized to GM-free, the indirect utility function is:

$$V^F = \frac{1}{p}(n + n^*)^{\frac{1-\theta}{\theta}} \tag{14}$$

If trade is permitted, and all goods are standardized to GM, the indirect utility function is:

$$V^G = \frac{\lambda}{p}(n + n^*)^{\frac{1-\theta}{\theta}} \tag{15}$$

Both harmonized indirect utility functions are graphed in Fig. 3.2.

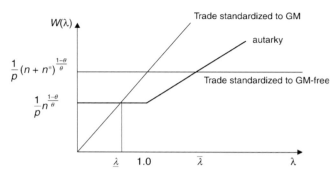

Fig. 3.2. Social welfare as a function of weight put on the consumption of the GM version: autarky and unilateral standardization.

In Fig. 3.2, $\underline{\lambda}$ is the solution to

$$\frac{\lambda}{p}(n+n^*)^{\frac{1-\theta}{\theta}} = \frac{1}{p}n^{\frac{1-\theta}{\theta}} \Rightarrow \underline{\lambda} = \left(\frac{n}{n+n^*}\right)^{\frac{1-\theta}{\theta}} \quad (16)$$

$\bar{\lambda}$ is the solution to

$$\frac{1}{p}(n+n^*)^{\frac{1-\theta}{\theta}} = \frac{\lambda}{p}n^{\frac{1-\theta}{\theta}} \Rightarrow \bar{\lambda} = \left(\frac{n}{n+n^*}\right)^{\frac{\theta-1}{\theta}} \quad (17)$$

Putting Fig. 3.2 into the EU–USA context, if the EU has $\lambda < \underline{\lambda}$, then for the EU, trade standardized to GM is worse than autarky. If the USA has $\lambda > \bar{\lambda}$, then trade standardized to GM-free is worse than autarky. Thus, whether the standard is GM or GM-free, in this case one country will be worse off than autarky and will not agree to the standard. If one country has a preference weight between $\underline{\lambda}$ and $\bar{\lambda}$, then standardization is possible. For example, suppose that the EU has a λ value between $\underline{\lambda}$ and 1. Then the EU would prefer trade standardized at GM-free to trade standardized at GM, which would be still preferred to autarky. The USA would prefer trade standardized at GM to autarky. Thus, in this case, trade standardized to GM is preferred to autarky by both countries.

When both countries have values of λ between $\underline{\lambda}$ and $\bar{\lambda}$, then trade with either GM or GM-free standards is preferred to autarky. However, the EU, with $\lambda < 1$, still prefers the GM-free standard to the GM standard. The USA, with $\underline{\lambda} > 1$, still prefers the GM standard. Which standard (if either) prevails depends on the bargaining game in place, and whether this bargaining game allows for a Pareto-improving equilibrium or suffers from a Prisoner's dilemma.

The possible outcomes depicted in Fig. 3.2 are particularly interesting when trying to explain the current trade impasse between the EU and the USA. One explanation is that the EU strongly prefers GM-free ($\lambda < \underline{\lambda}$), and that the USA strongly prefers GM ($\lambda > \bar{\lambda}$). In this case there is little hope for harmonization. The competing explanation is that both the USA and EU have values of λ between $\underline{\lambda}$ and $\bar{\lambda}$, but that they have been unable to solve the bargaining game. In this case, there is hope that continued negotiation – either directly or through WTO processes – will eventually result in a situation that is welfare-improving for both sides.

Suppose now that the EU and USA would allow trade in transgenic crops and GM foods without requiring harmonization of standards. The USA produces (and trades) only GM varieties (regardless of what the EU does). In that case, the indirect utility function from consuming both GM-free and GM made varieties (as earlier, assuming fixed and marginal costs are standard and variety invariant) is:

$$V = \frac{1}{p}\left(n+n^*\lambda^{\frac{\theta}{1-\theta}}\right)^{\frac{1-\theta}{\theta}} \quad (18)$$

where n is the number of varieties produced by the EU, and n^* is the number of varieties produced in the USA. It is possible that $n \neq n^*$ due to differing endowments across countries, not different technologies across countries. A representative consumer consumes a number of varieties previously available in autarky, as well as varieties produced in the other country. If $\theta > 0.5$, the 'non-standardized' trade social welfare is convex for $\lambda \in <0,1>$, and graphed in Fig. 3.3. If $\theta < 0.5$, the 'non-standardized' trade social welfare is concave for $\lambda \in <0,1>$, and graphed in Fig. 3.4. Non-harmonized trade is always welfare-increasing regardless of the weights in the linear utility function (λ).

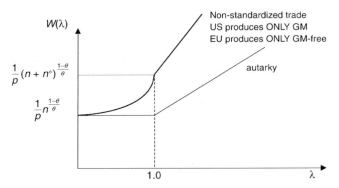

Fig. 3.3. Social welfare as a function of weight put on the consumption of the GM version: autarky and non-standardized trade, $\theta < 0.5$.

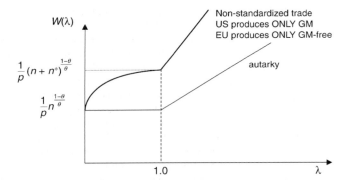

Fig. 3.4. Social welfare as a function of weight put on the consumption of the GM version: autarky and non-standardized trade, $\theta > 0.5$.

If we allowed for differences in marginal costs, it would change the 'critical value' (i.e. $\underline{\lambda}$ and $\bar{\lambda}$: where consumers are indifferent between GM or GM-free), and shift curves. Differences in country size or cross-country differences in fixed costs affect n and n^*.

We demonstrated that non-standardized trade in agricultural commodities between the EU and USA would be welfare-improving relative to autarky. The USA would be producing and exporting GM varieties, while the EU would be producing and trading GM-free varieties. On the other hand, if the marginal rate of substitution between GM and GM-free versions of the same variety is very low – as it might be in the case of GM in the EU – the country is better off in autarky than it would be with trade harmonized at the GM standard. Such harmonization of GM standards between the EU and USA can be 'vetoed' by the 'disadvantaged' country by not agreeing to the standards even if a larger number of varieties is available.

Now we extend the number of countries considered: assume some countries adopt the EU's precautionary approach to transgenic crops and GM foods (whether the precautionary regulatory approach was adopted in the 'follower' countries due to preference for GM-free or to secure access to the EU market is an interesting question but beyond the scope of this chapter), while some countries are as pro-technology as the USA.

If non-harmonized trade is not feasible – for example, for political reasons (i.e. one of the countries would attempt to negotiate a common binding position on the global level in the fear that non-harmonized trade would stop, or would result in a dispute settlement procedure) – the second best option is for countries to engage in trade with countries with similar preferences and similar regulatory approaches, and *de facto* form subglobal trade agreements to capture at least some of the gains from trade. The agreements do not necessarily have to be institutionalized in the WTO sense. This finding is consistent with the evidence on increasing numbers of small trade agreements of all types (not necessarily related to biotechnology), and fully supported by social choice and club theory (i.e. Buchanan, 1965; Olson and Zeckhauser, 1966; Olson, 1968; Cornes and Sandler, 1996). Groupings of countries with similar regulatory approaches become *de facto* clubs in a Buchanan sense, where standardized goods (or non-standardized goods, depending on whether precautionary principle or scientific equivalence was adopted) have the characteristics of club goods, and non-members are excluded from consuming the benefits, i.e. unrestrained trade among members. Workings of clubs are facilitated by their respective leaders: the EU in the 'GM-free club' and the USA in the 'GM club'. Leadership is facilitated by smaller, more homogeneous groupings of countries. The existence of clubs is also assisted by positive externalities for smaller countries of belonging to a certain club, such as economic networks, political allies, etc.

Unavoidably, the 'club' approach is accompanied by a number of dilemmas. Although the argument of securing welfare-maximizing gains from trade is a valid one, it comes at the expense of a separating equilibrium and lower welfare than non-harmonized trade. For a small country that has less to offer a large country (or club) in terms of gains from trade, there is a greater incentive to harmonize domestic standards with the large country standards, even if these standards are not what the country would have chosen without the trade considerations (e.g. the Zambia food aid case). In addition, the separation might reflect neither consumer preferences in respective countries nor the chosen regulatory approach to transgenic crops and GM foods. Rather, it might be representative of trade and the 'political' affiliation of countries. This hypothesis remains to be empirically tested – assuming a data set allowing such analysis would be available. As of now, GM and GM-free transgenic crops are treated as like products, and their major producers (the USA, Argentina and Brazil) do not have separate distribution channels to market their crops (with the exception of the organic channel and a few contract farming situations). Therefore, unless proven GM-free, all products are considered to be GM.

We expect to see increasing numbers of small countries adapting either EU-type standards, or neglecting food content standards in the hope of exporting to North America (and China and India

appear to be producing and importing GM foods). In contrast, Brazil seemed to be producing transgenic soybeans even in the face of a ban on the crop, and Argentina has quickly and widely adopted transgenic soybeans even without enforcement of intellectual property rights. The result, nevertheless, is a bifurcation of the world's agriculture into two clubs, one that accepts GM foods and one that does not, and a reorganization of trade and trade patterns based on this bifurcation.

Coexistence of GM and GM-free Crops

We showed that non-harmonized trade is always welfare-improving compared with autarky, whereas harmonized trade could be welfare-worsening. In the GM case, non-harmonized trade is equivalent to coexistence of GM and GM-free varieties on the same market. Recognizing that 'the ability to maintain different agricultural production systems is a prerequisite for providing a high degree of consumer choice,' in July 2003 the Commission of the European Communities (2003) recommended to the Member States (non-binding) 'guidelines for the development of national strategies and best practices to ensure the co-existence of genetically modified crops with conventional and organic farming' (henceforth Coexistence Guidelines). Coexistence Guidelines seek to establish that 'no form of agriculture, be it conventional, organic, or agriculture using GMOs, should be excluded in the European Union'.

Coexistence is defined as the ability of farmers to make a practical choice between conventional, organic and GM crop production, in compliance with the legal obligations for labelling and/or purity standards while addressing the most appropriate management measures that can be taken to minimize the admixture of GM and non-GM crops. The development and implementation of measures for coexistence is delegated to the Member States due to extremely diverse agricultural conditions across the EU. Management measures for coexistence adapted by states should reflect the best available scientific evidence on the probability and sources of admixture between GM and non-GM crops – permitting the cultivation of GM and non-GM crops, whilst ensuring that non-GM crops remain below the legal thresholds for labelling and purity standards with respect to GM food and feed, as defined by Community legislation.

The existing Guidelines, although leaving the details to be worked out by the member countries, are introducing a notion that as a general principle, during the phase of introduction of a new production type in a region, operators (farmers) who introduce the new production type should bear the responsibility of implementing the farm management measures necessary to limit gene flow.

The EU guidelines stirred a lot of attention – with some claiming that 'coexistence' in fact implies prohibition of biotechnology and transgenic crops and GM foods, and effectively acts as an agricultural policy tool protecting EU farmers.

The guidelines seem to reconcile protection of consumer preferences with the WTO's focus on producers. In the current WTO framework, consumer preferences *per se* are not recognized as a legitimate cause for protection. The coexistence guidelines allow for the production of GM crops, but actual production will take place only if producers and grocers feel they can sell GM foodstuffs to the consumer. Therefore, even with the Coexistence Guidelines in place, considering negative consumers' attitudes towards GM, and the additional burden placed on the farmers adopting GM, the dominant strategy for a farmer appears to be not to adopt GM (i.e. continue producing conventional or organic). Thus, the Coexistence Guidelines could have practical impact only if consumer acceptance of GM in the EU changed, providing farmers with a *bona fide* choice between conventional, organic and GM crop production.

apparently been lifted, with new varieties working through the approval process in early 2004. The EU is also exploring 'choice-based' ways in which regulation can allow the 'coexistence' of GM and GM-free varieties and foods. There is little evidence that these potential regulatory changes reflect changes in consumer response to GM foods (as opposed to pressure from the USA and the WTO).[11] Simultaneously, an increasing number of countries are faced with pressing needs to make choices regarding GM production and consumption. For example, several southern African countries faced the choice between malnutrition and possibly famine or accepting food aid in the form of GM maize, with the decisions being largely anti-GM. Other countries with large populations to feed – notably India and China – have formulated regulations that are expected to facilitate or at least allow rapid increases in GM area planted and GM food consumption. In other words, it appears that the technology is moving faster than the regulation. This suggests that at least for the immediate future, there will be an increasing number of countries who feel the urgency to make a choice regarding GM foods, leading to further bifurcation into 'pro-GM' and 'anti-GM' clubs.

Acknowledgements

Funding support from the Elton Smith in Food and Agricultural Policy at Michigan State University, and Michigan Agricultural Experiment Station, East Lansing, MI, is gratefully acknowledged. We thank Dr Steve Matusz for comments and suggestions on the model.

Conclusions

The opposing positions of the USA and the EU threaten to polarize the world's agriculture into a set of countries that accepts biotechnology and a set of countries that does not. Underlying this are different preferences regarding transgenic crops and GM foods, and different regulatory frameworks (precautionary principle versus scientific evidence of equivalence) built upon these preferences.

This polarization is becoming more noticeable despite some movement in the EU regulatory stance. The EU moratorium on the introduction of new varieties that was initiated in 1998 has

Notes

1 Although the EU member countries are not completely homogeneous in their positions towards GM, the EU as an entity endorses precautionary principle.
2 While it is yet to be established whether different regulatory approaches – both used in international treaties – are indeed *de jure* trade barriers, they are definitely *de facto* trade barriers.
3 For general and more technical exposition of the model, see Tothova (2004).
4 In the text, the words 'standard', 'quality', 'attribute', 'characteristics' and some specific examples are

used as synonyms, and shall be understood in a broader connotation.

[5] We take food in general grouping raw commodities and processed foods, while ignoring potential processing costs.

[6] As Monsanto suspended biotech wheat in May 2004 'bowing to global opposition', it is suspected that fear of ruining US dominance on the wheat market played a role. Foreign buyers announced they would not buy any US wheat at all if biotech wheat was approved because it might get mixed with conventional wheat supplies (http://money.cnn.com/2004/05/10/news/fortune500/monsanto.reut/index.htm?cnn=yes last viewed May 24, 2004). They also announced they are unwilling to buy the GM crop, not least because they see few benefits, whether for farmers or themselves (http://news.bbc.co.uk/2/hi/business/3702739.stm last viewed May 24, 2004).

[7] In the Helpman and Krugman (1985) terminology, actual varieties produced form a set of available varieties with finite price – varieties not available are considered to have an infinite price.

[8] We ignore integer issues when discussing number of firms in the economy.

[9] Assuming there is a large number of varieties so that each country produces a different and non-overlapping set of varieties.

[10] A third possible case is that the EU and USA might attempt to negotiate a common standard different from their current position, for example a common tolerance level that would weaken the EU's existing tolerance level, but be stricter than the 'present' US GM standard (which is non- existent, or practically corresponds to complete acceptance of transgenic crops and GM foods). However, as outlined earlier, 'compromise' of standards would imply a change in regulatory approaches in both the USA and EU (as would the unilateral harmonization). In practice, this would imply that the EU would depart from its (almost) zero tolerance policy, while the USA would admit the need to impose government or private regulation to police the industry and assist trade. Due to foreseen political difficulties, we will not explore this case.

[11] http//www.cnn.com/2004/TECH/science/04/28/bc.food.eu.gmo.reut/index.html

References

Bernauer, T. (2003) *Genes, Trade, and Regulation.* Princeton University Press, Princeton, New Jersey.

Buchanan, J.M. (1965) An economic theory of clubs. *Economica, New Series* 32, 1–14.

Commission of the European Communities (2003) Commission Recommendation on Guidelines for the Development of National Strategies and Best Practices to Ensure the Coexistence of Genetically Modified Crops with Conventional and Organic Farming. Available at: http://europa.eu.int/comm/agriculture/publi/reports/coexistence2/guide_en.pdf Last viewed September 10, 2004.

Cornes, R. and Sandler, T. (1996) *The Theory of Externalities, Public Goods, and Club Goods.* Cambridge University Press, Cambridge, UK.

Dixit, A. and Stiglitz, J. (1977) Monopolistic competition and optimum product diversity. *American Economic Review* 67, 297–308.

Helpman, E. and Krugman, P. (1985) *Market Structure and Foreign Trade.* MIT Press, Cambridge, Massachusetts.

Isaac, G.E. and Kerr W.A. (2003) Genetically modified organisms at the world trade organization: a harvest of trouble. *Journal of Trade Law* 37, 1083–1095.

Kerr, W.A. (2003) Science-based rules of trade – a mantra for some, an anathema for others. *Estey Journal of International Law and Trade Policy* 4, 86–97.

Krugman, P. (1979) Increasing returns, monopolistic competition, and international trade. *Journal of International Economics* 9, 469–479.

Krugman, P. (1980) Scale economies, product differentiation, and the pattern of trade. *American Economic Review* 70, 950–959.

Lusk, J.L., Roosen, J. and Fox, J.A. (2003) Demand for beef from cattle administered growth hormones or fed genetically modified corn: a comparison of consumers in France, Germany, the United Kingdom, and the United States. *American Journal of Agricultural Economics* 85, 16–29.

Moon, W. and Balasubramanian, S.K. (2003) Is there a market for genetically modified foods in Europe? Contingent valuation of GM and non-GM breakfast cereals in the United Kingdom. *AgBioForum* 6, 128–133.

Olson, M.J. (1968) *The Logic of Collective Action.* Schocken Books, New York.

Olson, M.J. and Zeckhauser, R. (1966) An economic theory of alliances. *Review of Economics and Statistics* 48, 266–279.

Perdikis, N. (2000) A conflict of legitimate concerns or pandering to vested interests? Conflicting attitudes toward the regulation of trade in genetically modified goods – the EU and the US. *Estey Centre Journal of International Law and Trade Policy* 1, 51–65.

Tothova, M. (2004) Truth vs. Beauty: Essays on Standards, Trade, and Agreements. Doctoral Dissertation. Michigan State University, East Lansing, Michigan.

Tothova, M. and Oehmke, J.F. (2004) Genetically modified food standards trade barriers: harmonization, compromise, and sub-global agreements. *Journal of Agriculture and Food Industrial Organization* 2, Article 5. Available at: http://www.bepress.com

World Trade Organization (2004) Various online materials. Available at: http://www.wto.org Accessed on May 7, 2004.

4 The Labelling of Genetically Modified Products in a Global Trading Environment

Stefania Scandizzo

Corporación Andina de Fomento, Caracas, Venezuela

One of the greatest technological developments of recent times has been the ability to genetically modify living organisms. This has brought great advances especially in agriculture, with the possibility of developing crops that are hardier, more resistant to pests and that generate greater yields.

As with all forms of technological development, there has been a backlash: ethical resistance to interfering with living creatures; the fear of loss of biodiversity; and the risk that genetic modification can pass to other organisms. Genetically modified (GM) food products have been generally available for quite some time now (although most consumers are not aware of it). However, two cases in Europe, the mad cow scare in the mid-1990s and the dioxide poisoning of many different types of food products in the summer of 1999, put the issue of GMOs (genetically modified organisms) on all the front pages and greatly increased the level of awareness and information of the average consumer. This has created a strong opposition to the presence of GMOs in food products, and the demand for greater information and decision power on the part of European consumers.

The issue of GMOs in food products may seem similar to that of organic produce, i.e. a case of quality differentiation that segments the market. There are, however, important differences. Organic produce is sold at a much higher price than equivalent traditionally produced goods, and targets a niche market of consumers willing to pay a premium for 'healthier' foods. Genetic modification instead tends to reduce the costs of production of many foodstuffs. Production in these products, however, tends to be extremely concentrated, and reduced costs are not passed on to the consumer. Furthermore, the characteristics of organic produce are highly advertised, while the presence of GMOs is usually neglected on food labels, creating an atmosphere of consumer distrust.

If cost savings from GMOs were passed on to consumers in the form of lower market prices, we could envisage a market where some consumers are willing to accept the uncertain health risks associated with GMOs for a lower price. The market would thus be segmented between GMO goods, traditional goods (which increasingly have a GMO component) and organic goods. However, since the price of traditional goods with GMO technology has not changed, there must be a different story going on.

Suppose consumers' preferences differ over different types of goods, with higher utility ascribed to goods without GMO presence. However, consumers are not able to discern GMO presence in goods, because at the marketing stage all varieties of the goods are combined into a unique good. Economies of scale in marketing are such that the combined good can sell at a lower price than the individual goods. An example of such a product is soybean production in the USA.

If labelling laws are enacted (such as those being presently discussed in the European Union (EU)), the marketing economies of scale are lost.

However, consumer utility may increase, as consumers are able to choose the variety that they most prefer.

In a closed economy, this is the end of the story. In an integrated world economy, things are more complicated. In a two-country world, use of GMO technology may be concentrated in one country, or at least unevenly divided. Labelling laws would therefore affect the two countries differently, with a possible positive effect for the country with a comparative disadvantage in GMO technology and instead a negative effect for the country exporting GMO goods. Labelling would therefore be seen by the latter country as a form of arbitrary trade barrier, in the same league with product and health standards. If labelling laws were decided unilaterally, we would expect the two countries to have different labelling criteria (less stringent in the GMO technology-abundant country), with the possibility of a trade war deriving from disagreement over differing criteria. If labelling laws were decided internationally, i.e. negotiated in the framework of a multilateral trade forum, the criteria would be much lower.[1]

In conditions of constant return to scale, the trade effects of labelling laws will be similar to those of international health or technology standards. In biotechnology products, however, as in many other food and agriculture products, there are economies of scale at the marketing stage. Labelling allows consumers to pick their preferred varieties, but increases prices. Producers of less preferred varieties will undisputedly lose; not only will they lose the economies of scale tied to joint marketing but they will see their demand decrease. Producers of more preferred varieties will face the same cost increase, but this may be outweighed by the increase in consumer demand for their specific product. In an international trading environment, the issue becomes one of profit shifting.

Little work has been done on the economic effects of labelling in the presence of economies of scale. Mattoo and Singh (1994) deal with the issue of eco-labelling, that has the objective of increasing consumer awareness about the ecological effects of products and thereby reducing production by environmentally unfriendly methods. Many authors, such as Hillman (1991), Mansfield and Busch (1995) and Chambers and Pick (1994), have dealt with the economic effects of various forms of non-tariff barriers to trade, which in some aspects can also apply to labelling. Finally, there is extensive literature on international trade in the presence of external economies of scale, such as Ethier (1982a) and Panagariya (1986). To include the effect of increasing returns to scale at the marketing stage, we will use the model developed by Ethier (1982b) which examined the interaction of national and international economies of scale in world trade.

The Model

The set-up of the model is a straightforward application of Ethier (1982b), which I shall follow closely in this chapter. Consider two goods, a traditional good and a biotechnology good. The latter is produced according to differing technologies by different individual firms, and subject to internal increasing returns to scale. In particular, some firms will adopt GMO technologies and others will not. All varieties of the product are then mixed and marketed as a single good. The marketing technology exhibits external economies of scale, i.e. cost reduction deriving from the number of varieties produced and accruing to all producing firms. These economies of scale are national in nature, i.e. 'marketing' is done for a specific country good. We can think of the GMO technology good as soybeans, and the two countries as the USA and Europe. All varieties of soybeans in each country are mixed and then sold as an individual country product: US soybeans versus European soybeans.[2]

Defined as W the traditional good; and M the biotech good (e.g. soybeans) a number $n > 0$ of different varieties of soybean are produced, some genetically modified, others not. x_i is defined as the amount produced of variety i of soybean.

All varieties of soybeans are mixed and marketed together. The marketing technology, which exhibits national increasing returns to scale, is given by:

$$M = n^{\alpha} \left[\sum_{i=1}^{n} \left(x_i^{\beta} / n \right) \right]^{1/\beta} \quad (1)$$

with $\alpha > 1$ and $0 < \beta < 1$. β is an index of substitutability: a higher value of β indicates that varieties of soybean are more easily substituted for each other in the supply of the final soybean product.

The effect of an increase in β is to lower the equilibrium price and increase the equilibrium output per brand. In a free entry Nash equilibrium, each firm must be maximizing profits but no operating firm makes positive profits. Therefore, if β increases, the number of varieties of soybeans in the free entry equilibrium falls. In fact, the more readily available a suitable substitute, the more intense is likely to be the competitive pressure on price.

While the traditional good is produced subject to constant economies of scale and a standard production function, the biotech good instead is produced subject to economies of scale (as shown above) and its production function is separable in the sense that it can be written as $M = km$, where k is an index of scale economies and m is an index of the scale of operations. While M is produced with increasing returns to scale, m is produced via a familiar smooth production function. Given factor endowment in the economy, this allows us to consider the relationship between W and M production as determined by a standard transformation curve:

$$W = T(m) \tag{2}$$

Another way of thinking of m is the amount of resources dedicated to the M sector.

For simplicity, assume all varieties are produced from factors of production via identical production functions, and that all varieties contribute symmetrically to the finished marketed product. Under these assumptions, all varieties will be produced in equal amounts and we can denote the output of each variety by x and the total amount of production as nx.[3]

Total costs for each firm producing soybeans, i.e. for x_i, are given by

$$C_i = cx_i + F_i \tag{3}$$

where F_i is the firm-specific fixed cost. An individual firm's production therefore exhibits traditional increasing returns to scale, completely internalized by each firm. Since we have interpreted the scale variable m as an index of the amount of factors devoted to sector M, we can suppose that the amount of factors needed to produce x units of any variety is $ax + b$, for some $a, b > 0$. Then

$$m = \Sigma a_i x_i + b_i \tag{4}$$

Assume for simplicity that there are two types of firms, firms that produce using GMO technology and non-GMO firms, and that type is decided before the analysis. Whatever fixed costs are associated with implementing GMO technology have been sunk, and at this point of the analysis a firm

cannot change its type. Independently of type, firms share an identical cost structure, although the resulting product is of differing quality.

Consider now consumer preferences. Consumers have preferences over different types of soybeans, but not over specific varieties. In particular, they have different preferences for goods with GMO content versus those free of GMO content, but they do not care about individual varieties of soybeans (in other words, they cannot distinguish over x). However, without labelling and given the marketing structure, they cannot distinguish between different types, and are forced to consume an aggregate product M. They would in fact prefer to consume a less aggregated aggregate product, M_1 (non-GMO product) and M_2 (GMO product) such that

$$M_1 = n_1^{\alpha}\left[\sum_{i=1}^{n_1}\left(x_i^{\beta}/n_1\right)\right]^{1/\beta}$$

$$M_2 = n_2^{\alpha}\left[\sum_{i=1}^{n_2}\left(x_i^{\beta}/n_2\right)\right]^{1/\beta} \tag{5}$$

where n_1 represents the number of firms producing GMO-free soybeans and n_2 the number of firms producing GMO soybeans.[4]

We can distinguish two cases.

Case 1: no labelling

With no labelling of GMO goods, maximum marketing economies of scale are achieved. This will be reflected in the price of the good. Consumers will benefit from lower prices, but will not be able to distinguish their consumption of GMO versus non-GMO goods. If different marginal utilities are associated with the two types of goods, this may not be optimal.

Case 2: labelling

Labels distinguish between GMO and non-GMO goods. In this case, we can still have marketing economies of scale but they will be lower: all GMO types will be bundled together, and all non-GMO types will be bundled together, and assuming identical cost structures across goods

$$M_1 = n_1^{\alpha}x$$

$$M_2 = n_2^{\alpha}x \tag{6}$$

and, given the economies of scale tied to marketing, $M > M_1 + M_2$

In this case, M will be smaller than in the case of no labelling, and therefore its price will be greater. (In particular, the price of M is inversely related to n.) This will have a negative effect on consumer welfare. However, the possibility of distinguishing between the two types of products lets consumers choose which product they prefer, which increases welfare. Which effect dominates will depend on consumer preferences, i.e. preference for non-GMO product, and measure of economies of scale.

Case 1: given that the consumer cannot distinguish between different types of soybeans, the consumer's problem is written in terms of the aggregate good M.

$$\max_{W,M} U = (1-\gamma)\log W + \gamma \log M$$

$$\text{s.t. } W + PM = Y \tag{7}$$

The solution to this problem is

$$W = (1-\gamma)Y \tag{8}$$

$$M = \gamma \frac{Y}{P} \tag{9}$$

Case 2: consumer's problem

$$\max_{W,M_1,M_2} U = (1-\gamma_1-\gamma_2)\log W$$
$$+ \gamma_1 f(M_1) + \gamma_2 f(M_2)$$

$$\text{s.t. } W + P_1 M_1 + P_2 M_2 = Y \tag{10}$$

where $M_1 + M_2 < M$

While previously the choice variable was M, now it is M_1 and M_2. If M_1 is the favoured good, the form of $f(M_1)$ becomes important, in that it determines the benefit of unbundling the goods for consumption.

From the production side, the story is as follows: there are n farmers who grow soybeans of different varieties. At harvest, soybeans are taken to a local buyer, i.e. a grain elevator, where farmers are given a price for their production. All soybeans, of all varieties, are then mixed together and sold as a single product to the final buyer, in this case a processor of final goods. By the time the good reaches the consumer, all varieties of soybean are indistinguishable. At the individual farm level, we have increasing returns to scale but a large number of farmers producing a differentiated product: this is monopolistic competition and will yield zero profits. Marketing has no costs except

those embodied in Equation 1, therefore the economies of scale are external to the firms and firms will behave competitively.

The price paid by the owner of the grain elevator to farmers is determined by cost minimization subject to Equation 1. If q_1 and q_2 denote the price, in terms of wheat, of two different varieties of soybeans, the cost minimization implies that production of each variety of soybean will be

$$x_1 = x_2(q_2/q_1)^{1/1-\beta} \tag{11}$$

If n is sufficiently large, i.e. if there are many different farmers being served by one grain elevator, the producer of each variety will act as though his behaviour does not influence that of other farmers. Then Equation 11 is the demand curve faced by farmer 1.

Each farmer must purchase the services of primary factors in competitive markets. (Remember that in terms of this model, primary factors are represented by m.) From our traditional transformation curve, the relative price of m in terms of W is $-T'(m)$, and the cost function of each individual farmer becomes

$$C_i = -T'(m)[a_i x_i + b_i] \tag{12}$$

Equating marginal revenue (which from Equation 11 is equal to βq_i) and marginal cost, each farmer will charge the price

$$q_i = -T'(m)a_i/\beta \tag{13}$$

The profit of each producing firm will be

$$\Pi_i = q_i x_i + T'(m)[a_i x_i + b_i] \tag{14}$$

Given the large number of firms and the *substitutability of varieties*, profits for all individual farmers will be driven to zero in equilibrium by the entry and exit of farmers (i.e. by variations in n). Therefore

$$x_i = \frac{b_i \beta}{a_i (1-\beta)} \tag{15}$$

Note that x_i is independent of n and of m, i.e. farmers' optimal choice of soybean production is independent of the number of competitors and of the amount of resources dedicated to the industry. Competition will drive out all higher cost firms, so that all producing firms will charge an identical price q_0.

Assuming zero profits, the relative supply price P_S of M in terms of wheat is given by equating the value of the finished product with the cost of production of the finished product:

$$P_S M = q_0 n x$$

$$P_S n^\alpha x = q_0 n x$$

$$P_S = n^{1-\alpha} q_0$$

M is produced with increasing returns to scale, therefore as n increases the supply price decreases (remember $\alpha > 0$). As noted previously, q_0 increases with P_S: this is intuitive, the higher the price obtained for the final product the higher the price of the intermediate goods.

On the demand side, assume that a constant fraction of income γ is always spent on M. Each output combination M and W determines a demand price P_D that will clear commodity markets:

$$P_D M = \gamma [W + P_D M] \tag{16}$$

Thus the demand curve can be written as

$$P_D = \frac{\gamma}{1-\gamma} \frac{T(m)}{n^\alpha x} \tag{17}$$

Note that P_D is also decreasing in n.

The price of the finished, marketed product, M, depends on the number of varieties bundled together. The quantity produced of each individual variety is independent of the marketing technology, of the number of varieties present and how they are bundled together. The price of each individual variety is increasing in the price of the bundled good, and therefore decreasing in the number of varieties bundled together. This is intuitive: an individual producer gains from sharing the expenses of marketing with other producers. (*Unless*, they think they have a variety preferred by consumers, and then they will wish it to be labelled as different and forego the joint marketing benefits.)

Labelling

What is the effect of labelling in this scenario? Labelling could potentially introduce an extra cost at the marketing stage, but this is not its most interesting effect, and we will not focus on it here. The issue of labelling becomes interesting when consumers present differing tastes over different types of the primary good, and labelling laws are enacted that respond to these differing tastes. In this case, labelling will affect how firms decide to market their goods, in this case, how many (and more importantly, which) different varieties will be marketed together.

From the point of view of production, labelling reduces the number of varieties that can be bundled together as one final product. As assumed above, there is no explicit cost of labelling. Furthermore, it does not affect the quantities produced or sold *by each farmer*, as we saw that the quantities of the primary good produced are independent of price or number of competitors, and depend only on technological parameters and the index of substitutability.[5]

In the case of differing consumer preferences across types, labelling will create a separate demand for certain subsets of varieties. Producers of more preferred varieties will face a trade-off: marketing oneself as the preferred type allows them to command a higher price, but by deciding to not market together with less preferred varieties their total costs increase. Producers of less preferred varieties are undeniably worse off: not only will consumers be willing to buy only for a lower price, but they will lose economies of scale in marketing since the more preferred varieties will not wish to share marketing costs.

From the point of view of consumers, reducing the number of varieties that can be bundled together increases the price of the final product, which is welfare reducing. If consumers are indifferent over different varieties of the intermediate product, this is the end of the story, and labelling will always be a welfare-reducing policy. However, if consumers have a preference over varieties or sets of varieties, and the labelling and bundling rules reflect this preference, labelling can have a welfare-increasing effect by allowing consumers to choose their favourite. The net effect will depend on parameter values, in particular the degree of consumer substitutability between the newly labelled products and the specific functional form of utility.

We must be specific about the labelling legislation. Assume that the legislation is such that all producers *must* label whether their good has GMO content or not. In other words, labelling is mandatory (like nutritional labelling), e.g. a red box for GMO goods and a blue box for non-GMO goods. Alternatively, we could assume that producers *must* label GMO presence in their goods, but they are not *required* to label *non*-GMO presence. Following our example, a red box is mandatory for

GMO goods while non-GMO goods can decide whether they want to use a blue box that distinguishes them from their GMO counterparts or a red box by which they cannot be distinguished. This is in fact the essence of currently discussed labelling laws. It may seem a subtle distinction, but it can have the following effect. Even in the case of consumer preference for certain varieties, if this preference is not sufficiently pronounced, the producers of preferred varieties may choose not to differentiate their products, irrespective of the labelling laws, i.e. if the losses in marketing economies of scale outweigh the price effect due to consumer bias, the producers of preferred varieties will choose to continue to bundle their products with other producers, and labelling will have no economic effect.

Welfare effects of labelling

To examine the welfare effects of labelling, there are four potential effects to be taken into account: (i) consumer benefit from being able to distinguish and choose preferred varieties; (ii) consumer loss from higher prices; (iii) effect on producers of preferred varieties; and (iv) effect on producers of less preferred varieties.

To consider the effect of labelling on consumers, we must first specify a utility function. I choose a Cobb–Douglas specification since it is the easiest to work with, and has the convenient characteristic of constant income shares. In this case, this seems like a reasonable assumption to keep: with the increased information due to labelling, consumers should not want to spend their income differently across goods, just choose the varieties they prefer for each good. Maintaining the Cobb–Douglas specification, we can have different effects of labelling. Assume that the income share devoted to the biotech good stays constant, and is divided between GMO goods and non-GMO goods, with $\gamma_1 + \gamma_2 = \gamma$. Furthermore, assume an added benefit to consumers from consuming the preferred variety, in our case M_1, denoted as δ.

$$U = (1-\gamma)\log W + \gamma_1 \log(\delta M_1) + \gamma_2 \log M_2 \quad (18)$$

Since this can be rewritten as

$$U = (1-\gamma)\log W + \gamma_1 \log M_1$$
$$+ \gamma_2 \log M_2 + \gamma_1 \log \delta \quad (19)$$

the income shares will remain constant.

Consumer effect

Regarding demand price, we must look at two effects deriving from labelling: the economies of scale effect (embodied by n) and the preference effect (embodied by γ, the fraction of income devoted to soybeans). Regarding the latter, labelling allows consumers to distinguish between more favoured and less favoured varieties, so

$$\gamma_1 + \gamma_2 = \gamma \quad (20)$$

where γ_1 is the share of consumer income dedicated to non-GMO soybeans and γ_2 the share dedicated to GMO soybeans. New demand prices will become

$$P_{D1} = \frac{\gamma_1}{1-\gamma_1-\gamma_2}\frac{W}{M_1} = \frac{\gamma_1}{1-\gamma_1-\gamma_2}\frac{T(m)}{n_1^\alpha x} \quad (21)$$

and

$$P_{D2} = \frac{\gamma_2}{1-\gamma_1-\gamma_2}\frac{W}{M_2} = \frac{\gamma_2}{1-\gamma_1-\gamma_2}\frac{T(m)}{n_2^\alpha x} \quad (22)$$

where all 1 subscripts refer to non-GMO goods and 2 subscripts to GMO goods.

The difference between the new demand prices P_{D1} and P_{D2} and with respect to the initial demand price P_D depends on the relative weights of consumer preferences (γ_1 and γ_2) for the different varieties and the relative number of firms producing the different varieties (n_1 and n_2). The greater the preference for a certain variety, and the lower the number of firms producing it, the greater will be the price. If instead the relationship between the ratio of preferences for non-GMO goods versus GMO goods and the ratio between the number of firms producing the two varieties were such that

$$\frac{\gamma_1}{\gamma_2} = \left(\frac{n_1}{n_2}\right)^\alpha \quad (23)$$

there would be no difference in demand price of the two varieties, but in any case the price would be greater than in the no labelling case (because of the loss of economies of scale):

$$P_{D1} = P_{D2} > P_D \quad (24)$$

Note this is a very particular case!

Another possibility is that $P_{D1} = P_D$ if

$$\frac{\gamma_1}{\gamma} = \left(\frac{n_1}{n}\right)^\alpha \quad (25)$$

but then $P_{D2} > P_D$. The opposite is also true.

In general $P_{Di} > P_D$ if:

$$\frac{\gamma_i}{\gamma} > \left(\frac{n_i}{n}\right)^\alpha \qquad (26)$$

It is possible for the price of each type to be greater than the price of the combined good, or for the price of one good to be greater than or equal to that of the combined good while the other is lower, but it will never be the case that both types of good have a lower equilibrium price than the combined good.

Consumers will prefer a labelling equilibrium to a non-labelling equilibrium only if there is a pronounced preference for one type of soybeans and the economies of scale effect on prices is not too pronounced. If instead consumers are relatively indifferent between types ($\gamma_1 \approx \gamma_2 \approx 0.5$), a non-labelling equilibrium will be preferred to be able to take advantage of the economies of scale effect. In fact, the utility from not labelling is

$$U(NL) = (1 - \gamma)\log W + \gamma \log M$$
$$= \log Y - \gamma \log P + (1 - \gamma)\log(1 - \gamma)$$
$$+ \gamma \log \gamma \qquad (27)$$

while instead the utility from labelling is

$$U(L) = (1 - \gamma)\log W + \gamma_1 \log M_1 + \gamma_2 \log M_2$$
$$= \log Y - \gamma_1 \log P_1 - \gamma_2 \log P_2$$
$$+ (1 - \gamma)\log(1 - \gamma) + \gamma_1 \log \gamma_1 + \gamma_2 \log \gamma_2 \qquad (28)$$

Producer effect

Once we have introduced labelling, what happens to firms' profits given what we have assumed about market structure? The assumption was that given the large number of firms and the substitutability of varieties, profits for all individual firms would be driven to zero in equilibrium. So competition drives out all higher cost firms, and all producing firms will charge an identical price.

On the production side, the effect of labelling is essentially to reorganize the market structure of the industry. Labelling changes the price individual farmers can command, and reduces the economies of scale for the local buyer/grain elevator.

As seen above, individual farmer's price is directly related to the price of the final product: $dq/dp > 0$. Since individual production levels are independent of price, a firm's profit effect will depend on whether the final demand price increases, decreases or remains constant.

Therefore, firms will prefer to separate as long as the profit effect is positive, i.e. as long as the price effect outweighs the economies of scale effect. This will occur when γ_i is high, i.e. when the preference for the produced variety is high, and when n_i is large enough that the economies of scale lost are not too large.

In a closed economy, that is the end of the story: either we have a separating equilibrium, with consumers and non-GMO producers better off and GMO producers worse off, or a pooling equilibrium, identical to the no labelling equilibrium. Note that since it is possible that one equilibrium is superior to the other in terms of social welfare, it will then be up to the government (or whatever authority imposed labelling) to keep this in mind in drafting the labelling legislation. For example, from the producers' point of view it may be preferable to pool, but for consumers the separating equilibrium is better, and consumer gain from labelling outweighs producers' net loss. In this case, it would be optimal for the government to impose a labelling law that requires all firms to specify their type.

The next question is what happens in a trading environment?

Labelling in a trading environment

Assume consumer preferences are the same across countries.[6] Consider first the case of a pooling equilibrium in both countries, which is equivalent to an equilibrium without labelling. This will result in inter-industry trade between the two countries: the country with a comparative advantage in production of the biotech good will be the net exporter.

Consider now a separating equilibrium in both countries, such that would result from both countries imposing labelling laws. If GMO and non-GMO producers are distributed equally across the two countries (i.e. the relative number of firms is the same), the producer effects will be the same in the two countries. Welfare effects will instead be skewed if one country has a relative abundance of GMO producers and the other a relative abundance of non-GMO producers. The country with more non-GMO producers will have a comparative advantage in the production of that type of good, will be able to offer a lower price and therefore will be the net exporter of the type. The country with more GMO producers will have a comparative advantage in that type of good, even though it is less preferred, and will become the net exporter in the relatively smaller market. We will witness two-way, or intra-industry, trade in biotech goods.

In the long run, if firms can change their production technology (whether they produce GMO goods or not), the comparative advantage of each country will become more entrenched. Given that marketing economies of scale are external to the firm but national in scope, if the majority of the countries' firms produce a certain type of good, it will be in the best interest of each firm to produce that type. In this sense, history has a role in determining the final outcome.[7]

What happens if the government in one country decides to impose mandatory labelling, while its trading partner does not? Firms in the other country will then be forced to separate if they wish to cater to the export market. Given the dimension of the price and economies of scale effect, it might be preferable to renounce the export market. In any case, this will imply a loss of revenue to exporters, and could result in a trade dispute, where the labelling law is seen as an indirect form of protectionism.

Even if the separating equilibrium is not mandatory in one country, the equilibrium will change in a trading environment with one country separating and the other pooling. Consumers will be able to choose between more expensive 'labelled' varieties, and the cheaper 'mixed' variety. Consumers will be able to choose between the preferred variety, at a higher price, or the less preferred variety, at a lower price. Depending on consumer preferences, and assuming identical preferences across consumers and countries, one variety may be priced out of the market. Over time, production in the separating country will be shifted into the more preferred variety, and the country's comparative advantage in this variety will increase, while the less preferred variety will disappear.

Concluding Remarks

The debate over GMOs has also spread to the area of international trade. Labelling laws have been seen as a way of resolving the issue by putting more power into the hands of consumers. However, given the asymmetric welfare effects of such laws, trade disputes have resulted. In an international trading environment, labelling laws not only have welfare effects in the short run, but, in the presence of economies of scale, will also have long-term effects on production and trading patterns. This is particularly relevant in the case of GM agricultural goods where economies of scale are strong.

Until now, the issue of labelling of GMOs has been restricted to trade between the USA and Europe. However, given the growing importance of GM crops in developing countries, how this issue is resolved in their developing trading partners could have relevant repercussions in these countries as well.

Notes

[1] See Miller and VanDoren (2000) for a discussion on labelling versus market solutions to separating goods with and without GMO presence.

[2] In Ethier (1982b), wheat is produced subject to constant returns to scale while manufactures are subject to increasing returns to scale. Finished manufactures are costlessly assembled from intermediate components. Distinctions among intermediate goods are not considered, which instead are the focus of discussion in this chapter.

[3] In reality, we would expect GMO crops to entail lower costs than their traditional counterparts. We will ignore this aspect for now, and assume – as will be explained later – that firm type (i.e. whether it produces GMO crops or non-GMO crops) is pre-determined.

[4] From here on, the subscript 1 will always refer to non-GMO variables and the subscript 2 to GMO variables.

[5] The index of substitutability, β, does not change because it represents substitutability within each bundled subgroup.

[6] Equivalently, we can assume there is no internal market for soybeans in either country and that all production is destined for a third market.

[7] Krugman (1991) considers not only the effect of history on determining the long run outcome in the presence but also that of self-fulfilling expectations.

References

Beath, J. and Katsoulacos, Y. (1991) *The Economic Theory of Product Differentiation*. Cambridge University Press.

Caswell, J. (1998) How labeling of safety and process attributes affects markets for food. *Agriculture and Resource Economics Review* 27, 151–158.

Chambers, R. and Pick, D. (1994) Marketing orders as nontariff trade barriers. *American Journal of Agricultural Economics* 76, 47–54.

Dixit, A. and Stiglitz, J. (1977) Monopolistic competition and optimum product diversity. *American Economic Review* 67, 297–308.

Ethier, W. (1982a) Decreasing Costs in international trade and Frank Graham's argument for protection. *Econometrica* 50, 1243–1268.

Ethier, W. (1982b) National and international returns to scale in the modern theory of international trade. *American Economic Review* 72, 389–405.

Hillman, J. (1991) *Technical Barriers to Agricultural Trade.* Westview Press, Boulder, Colorado.

Krugman, P. (1991) History vs. expectations. *Quarterly Journal of Economics* 106, 651–667.

Mansfield, E. and Busch, M. (1995) The political economy of nontariff barriers: a cross-national analysis. *International Organization* 49, 723–749.

Mattoo, A. and Singh, H. (1994) Eco-labelling: policy considerations. *Kyklos* 47, 53–65.

Miller, H. and VanDoren, P. (2000) Food risks and labeling controversies. *Regulation* 23, 35–39.

Panagariya, A. (1986) Increasing returns dynamic stability, and international trade. *Journal of International Economics* 20, 43–63.

5 Tree Biotechnology: Regulation and International Trade

Roger A. Sedjo

Resources for the Future, Washington, DC, USA

Introduction and Background

The advent of planted forests has provided a financial incentive to undertake tree improvement, since the benefits from investments in tree improvements can be captured in the higher productivity at harvest associated with the improvements. Tree improvement in industrial forestry has been underway in a significant way for about half a century, or roughly since the period when planted forests for industrial wood production began to become common. Increased productivity has resulted from the planting of exotic species, many of which have far greater yields in their new locations. Additionally, tree improvements to both indigenous and exotic trees have been brought about through activities such as superior tree selection, traditional breeding techniques and clonal propagation of superior trees. These innovations have had a positive influence on industrial wood production, particularly in some regions, thereby impacting on regional and international patterns of forest resource production and forest products trade.

In the USA, plantations have been a factor in the shifting of the centre of forestry from the west to the south. Globally, most notable is the growing role that South America is playing in the production of plantation-grown industrial wood and the export of wood and wood products. However, other regions have also participated, including New Zealand, Australia and South Africa.

Thus far, neither biotechnological regulations nor international trading rules have seriously impacted these shifts. Although international trade markets do differentiate among species, they do not differentiate between indigenous and exotic wood. Nor has international trading of the tree improvement innovations of the type undertaken been seriously restricted by trade regulations on either the wood product or the traditionally improved seed. However, aspects of transgenic (bioengineered) trees are likely to be treated quite differentially by international trading and existing regulatory regimes.

In most countries, transgenics, unlike most traditionally modified plants, are automatically regulated and required to pass successfully through a deregulation process for commercialization to occur. However, the deregulation criteria and processes often differ substantially among countries. For example, the acceptable level of risk associated with the deregulation of a plant transgenic is significantly different in the USA from what it is, for example, in the European Union (EU) (Pachico, 2003; Sedjo, 2004a). Nevertheless, currently, many transgenic agricultural crops have successfully completed deregulation and are integrated into domestic and international agricultural markets. However, disputes continue as to whether and under what conditions countries can refuse to import transgenic crops and prohibit the use of transgenic seed, e.g. within the context of the World Trade Organization (WTO).

In this chapter, I discuss issues and problems related to international trade of transgenic wood and of transgenic tree seed.

Concerns about Transgenics

Concerns about plant transgenics usually fall into one of two categories: health and safety; and environmental concerns. Health and safety generally relates to the consumption of the transgenic plant by humans or animals and any deleterious effects the transgenic may have on human or animal welfare. In addition, there are concerns about the effects of the transgenic plant on the natural environment. These include toxins that may flow from the plant as well as problems with the transgenic plant becoming a pest or an invasive. Also, there are concerns with 'gene escape' whereby the transferred gene might escape to a wild relative thereby increasing the fitness of that relative, thus enhancing its ability to become a pest. Finally, there are concerns that an escaped gene might despoil a pristine species collection and thereby compromise its usefulness for developing improved hybrids of a particular plant, e.g. maize. It is this set of concerns that has resulted in the automatic regulation of a transgenic plant and the requirement that for commercialization it must successfully pass through a deregulation process that assesses and examines the risks of any adverse or damaging effects that could be associated with its deregulation.

Outputs of Tree Biotechnology and Inhibitors to Transgenic Development

There are two separate and distinct products that can be the outgrowth of a transgenic tree. The first one is the transgenic wood that has been produced. The second is the transgenic seed or germplasm, which embodies the desired gene. Both of these can be traded on world markets; however, the trading characteristics differ since one of the outputs, the germplasm, is regulated, while the other, the wood, is probably not regulated in most cases.

Tree improvement for wood production, which has been undertaken over the past 50 years, largely involves the application of traditional (sexual) breeding techniques and the use of clonal propagation for some deciduous trees, e.g. poplar

and eucalyptus. Also, traditional techniques have generated a wide array of innovations particularly in orchard and decorative trees. The major emphasis of traditional breeding has been toward growth and yield, while maintaining tree form.

The advent of tree genetic engineering, which involves the asexual transfer of genes, has been in the research phase for only a few years and has tended to follow the experience of transgenic crops. For example, much early work on tree genetic engineering focused on the transfer of a herbicide resistance gene, in a fashion akin to that in some agricultural crops, e.g. soybean (Sedjo, 2004b). The more recent focus of biotechnology research, however, has been on modifying wood fibre characteristics. The conventional wisdom today is that traditional breeding approaches will be used to achieve increased growth and biomass yields, while genetic engineering will focus on obtaining the desired wood characteristics. Desired characteristics are those that deal with fibre availability and/or quality. These would include an increase in the amount of useful fibre or the production of fibre that is more cheaply processed in the digester in the production of wood pulp, thus increasing output yields.

Other research has been undertaken to develop a Bt gene, which provides natural protection against certain pests. However, these activities seem to be attracting less research attention recently. The reasons seem to be found in both regulation and in market size. At this time, there is limited research on tree Bt genes because, among other considerations, deregulation in the USA would be required by two separate agencies, the Animal and Plant Health Inspection Services (APHIS) under the plant protection act, and the Environmental Protection Agency (EPA), under its regulatory responsibilities for toxic substances. Since deregulation is costly, innovations that require two deregulation processes are to be avoided. Additionally, market size is an important consideration for research development. Trees are not an annual crop and thus only a small portion of the forest is replanted in any single year. A 20-year rotation, for example, requires tree replacement on an average of only 5% of the land. Hence, the market is small, when compared with some annual crops.

In a world of many countries with varying requirements for both deregulation and trade, deregulation in one country, e.g. the USA, does not ensure that the deregulated transgenic product can be marketed in other countries. Restrictions could be found in local deregulation requirements and/or in trade restrictions related to transgenics. Many countries require their own deregulation process, which might be required even if the transgenic has been deregulated in another country. This additional deregulation adds to the costs. Furthermore, for countries with small domestic markets, the benefits of the introduction of the innovation may not justify the costs of deregulation in that country. Chile, for example, has an interest in the herbicide-resistant gene, but apparently the US developer does not view the market as sufficiently attractive to justify discussion about its adaptation to Chile's planted *Pinus radiata* forests.

Given considerations of deregulation costs, additional deregulation requirements in second countries, possible trade restrictions and small markets, developers seem to be focusing on innovations which could achieve deregulation in products and species oriented toward relatively large markets. For example, a transgenic tree development firm in the USA is focusing on innovations for loblolly pine, the dominant plantation timber tree in the USA. Furthermore, large areas of loblolly pine are being established in several countries of South America, so the product would have potential in several foreign markets should their regulatory hurdles be negotiated.

Another major tree genus under major transgenic development is the eucalyptus, with major innovations apparently underway oriented towards production in Brazil. Interestingly, until recently, Brazil prohibited transgenics. However, it was acknowledged that transgenics were widely planted in certain annual crops, and the Brazilian firms apparently invested substantially in transgenic tree development even while the prohibition was in force. The assumption is that the prohibition is seen as very likely to be removed soon. Eucalyptus is widely and successfully planted in Brazil and there is surely a large domestic market. Also, worldwide, eucalyptus is one of the mostly widely planted wood trees used for pulp and also timber.

The State of Transgenic Wood and Tree Germplasm Activities Today

With the exception of one orchard tree, the papaya, no transgenic trees are known to be officially deregulated. However, informal sources have stated that transgenic trees have been deregulated and planted for commercial purposes in China.[1] However, there are a number of situations in which a transgenic tree is close to commercialization. As noted, China may already have commercial transgenics planted. Also, Brazil appears very near to the commercialization of a transgenic eucalyptus. The intent in Brazil is to introduce a transgenic tree in the next year or two that would use the very rapidly growing trees now in use as a platform for the insertion of genes that will dramatically increase (+40%) the fibre content of the tree. It is interesting that when this project was begun, Brazil still had a prohibition against the use of plant transgenics. However, that is being changed and commercialization of the tree is anticipated within a very few years. Also, New Zealand has developed a transgenic radiata pine, which is awaiting deregulation. However, new deregulations have been put on hold in New Zealand.

Finally, large investments in the development of a transgenic loblolly pine are underway in North America. The innovations would largely involve fibre modifications with the intention of improving the pulping characteristics of pine fibre, thereby lowering pulping costs. The technical challenges involve: first, the development of an appropriate mechanism to transfer gene(s) that generate the types of desired fibre modifications; and secondly, the development of a technique that would allow the massive replication required for plantation establishment in a low cost and timely manner.

International Trade of Wood and Transgenic Tree Germplasm

A simple model of international trade predicts goods will flow from the country of comparative advantage (low opportunity cost) to other countries with higher costs (Sedjo, 2004c). In the case of international trade in wood, the basic product is raw wood, but availability of raw wood often allows a range of products to be produced at low cost, from building materials, to pulp and paper, and also may include a variety of wood products and would include packaging materials of both paper and wood. The complex variety of products, as well as the benign nature of transgenic wood,[2] makes it unlikely that any serious prohibitions to the international trade would be constructed as the result of transgenic wood.

However, the transmission of tree germplasm in the form of seed or seedlings could be viewed as a very different event. For trees, the concerns expressed might be largely oriented toward the impact of the planted tree on the natural environment through gene escape (Williams, 2004).

In fact, the concerns might vary by region. It is generally agreed that little gene transfer is likely to occur in the natural environment when very different species are involved (DiFazio *et al.*, 1999). Thus, for example, since pine is not indigenous to South America, the probability of gene transfer to indigenous tree species, which do not include pine, is non-existent. The same argument is applied to eucalyptus, an exotic, in North America. Thus, concerns related to the international trading of germplasm are likely to vary by the conditions particular to the region, particularly the species composition.

One question that has been raised is the extent to which tree improvements and improved transgenic trees are likely to generate major shifts in the comparative advantages of the various regions. In fact, with the advent of planted forests over the past 50 years, the world has already seen a major restructuring among the leading regional timber producers. First, exotics have been widely planted in regions where they are particularly suited. Secondly, intensive management is increasingly being practised, especially on highly suitable sites. Finally, tree improvement programmes further improved the growth and yields, again particularly in regions well suited for planted forests.

As a result of these forces, wood harvested from planted forests has increased from a negligible fraction of total harvests to roughly 34% today (Food and Agriculture Organization, 2001). These increases have occurred both in traditional timber-producing regions, e.g. the US South, and in newly established planted regions, e.g. South America and New Zealand. There is every reason to expect these shifts to continue, and indeed be facilitated by transgenic forestry.

Obviously, transgenic forestry also has the potential to accelerate some of these trends. However, it could also modify them. For example, to the extent that transgenic forestry can improve tree performance in temperate and boreal sites, it may be able to facilitate a resurrection in lightly inhabited northern regions. Relatively rapidly growing trees on low opportunity cost northern sites could improve the competitive position in these areas of otherwise low economic activity.[3]

Barriers to Trade in Transgenic Trees and Wood

Potential barriers to transgenic forestry and wood arise at two levels. First would be barriers to transgenic wood flows. These would seem to be unlikely due to the benign nature of transgenic wood, as discussed above.

Secondly would be barriers to the flow of tree germplasm. These barriers are likely to be formidable. Individual countries have deregulation standards and procedures. Although countries may accept the concept of reciprocity, honouring the deregulation process of others and thus accepting live transgenic tree germplasm, this may well not become the rule. In many cases, individual countries may want to deregulate a transgenic tree before it is allowed to be planted commercially. Since the deregulation systems will vary by country, a transgenic tree that is deregulated in one country may not be recognized in other countries.[4] Such a system would be fraught with uncertainty and quite costly. In many cases, it may simply not be worth the cost to deregulate a tree in a country with a modest growing potential.

Overcoming Barriers to Transgenic Wood Trade

Given the wide degree of interest in transgenic trees, it appears likely that some transgenic wood trees will eventually become commercialized in some countries of the world. Thus far, one transgenic tree has been 'officially' deregulated in the world, specifically the orchard tree, papaya, which was deregulated by the USA (by APHIS) and is now commonly planted in Hawaiian orchards. It may, however, not be in the major producing countries, but rather in countries of South America or Asia, in which major deregulation takes place for timber trees. It has been unofficially reported, for example, that some 300 ha of transgenic poplar have been established in a commercial forest in China.[5]

Based on some of the considerations discussed above, it appears that the wood of transgenic trees is unlikely to be effectively excluded from international trade. The wood can take many forms, would be difficult to detect and, since it appears to be harmless, appears not to have characteristics that would warrant serious control costs. Thus, it appears that the decision as to whether to commercialize will depend on the basic financial returns to the innovation, the degree of restriction and the costs of deregulation in individual countries. It seems likely that some countries will deregulate and allow the production, sale and export of transgenic wood. Other countries may not allow production, but are likely to find it difficult to prohibit importation of transgenic wood and especially wood and paper products. Thus, it appears likely that the market will eventually find transgenic and non-transgenic wood competing in various wood markets.

Forest Certification as a Barrier to Transgenic Wood

In recent years, there has been a strong movement toward forest certification to ensure that wood is managed in a sustainable or 'well managed' manner. There are a number of organizations that have created standards, sponsor forest audits and represent themselves as 'certifiers' of the management of commercial forests. At least one of the major certifying groups, the Forest Stewardship Council (FSC), has standards that will not allow certification of a forest that contains transgenic trees. Of course, at this point in time, such a requirement is moot since there are essentially no deregulated transgenics to plant. However, the FSC has gone so far as to withhold recognition from forest firms that have research activities related to the development of transgenic trees.

How effective using certification as an approach to preventing the commercialization of transgenic trees will be in preventing the development and utilization of transgenics remains to be seen. While FSC is the only forest certifier active in most parts of the globe, there are a number of other major forest certifiers, often regionalized but usually in competition with at least one other certifier (e.g. see Cashore and Lawson 2003).[6] Some of the major certifiers are more accepting of

transgenics, requiring only that the tree planters follow the existing law and practice sound science (e.g. SFI). It is feasible that a strong public preference for certified wood and a preference for the certification of the FSC could inhibit the development of a transgenic wood market. However, there is little evidence currently that the preference for certified wood translates into higher prices as sometimes alleged (Sedjo and Swallow, 2002). The lack of a price premium may reflect a relatively weak overall preference for certified wood, which may or may not transfer to transgenic wood.[7]

Implications of Barriers for Transgenic Trees

It appears unlikely that any of these barriers, with the possible exception of certification, will provide a significant obstacle to the free flow of transgenic wood. However, several of these barriers are likely to provide a substantial obstacle to the international flow of transgenic tree germplasm. This would result in a world where the good (wood) would be traded but important technology (transgenic trees) would not. To the extent that transgenic trees provide low cost and perhaps superior wood, some regions will have a strong financial incentive to adopt the technology. This could provide an additional comparative advantage to the countries and firms that adopt transgenic trees.

Furthermore, to the extent that countries that already have a comparative advantage in wood production benefit disproportionately[8] from transgenic trees and therefore would adopt the appropriate technology (transgenic trees), and countries without a comparative advantage fail to adopt transgenics, the comparative advantage gap would widen. Thus, one could envisage a world in which wood production specialization becomes even more intense, with a few high wood productivity countries, which have adopted the latest transgenic technology, further increasing their share of worldwide timber production.

Summary and Conclusions

This chapter suggests that for transgenic trees to have an opportunity to have a major impact on global production requires that they be developed, deregulated and commercialized. It discusses the regulatory barriers to commercialization, which may vary by country, and examines the potential implications of these barriers on international trade in transgenic wood and in transgenic tree germplasm. Development is being undertaken in many countries of the world with the focus apparently on improving wood fibre. Since deregulation is undertaken country by country, without the acceptance of some countries of the deregulation undertaken in others, the international transfer of transgenic tree germplasm could be stifled. There is little reason to believe that transgenic wood flows would be negatively impacted unless there was widespread opposition that translated itself into prohibiting certification standards for transgenics, which were widely accepted. However, transgenic trees offer the possibility of increasing country specialization by enhancing the comparative advantage of existing exporters, while importers fail to use the technology due to lack of local deregulation. This could result in increasing the regional degree of specialization in wood production and result in larger volumes of exports from increasingly dominant global wood producers.

Notes

[1] China has a process for the deregulation of transgenic plants (Pachico, 2003).

[2] For example, wood used for pulp, paper and materials is dead matter, and not edible by humans.

[3] In fact, genetically improved trees (although not transgenics) are currently being planted in Minnesota and central Alberta in an attempt to provide additional timber resources for current ongoing resource development activities. In Alberta, planted forests are anticipated to compete away land formerly in crop agriculture.

[4] This appears to be the type of system often practised in agriculture today.

[5] At a November 2003 meeting of the FAO Panel of Experts on Forest Gene Resources, a Principal Research Scientist of the Research Institute of Forestry in China reported on the establishment of close to 300 ha of commercial transgenic poplar in China. (Yousry El-Kassaby, personal communication, January 20, 2004).

[6] The major forest certifiers are: FSC, Forest Stewardship Council; SFI, Sustainable Forest Initiative; CSA, Canadian Standards Association; ATFS, American Tree Farm System; PEFC, Pan European Forest Certification; MTCC, Malaysian Timber Certification Council; LEI, Lembaga Ekolabe Indonesia; ISO, International Standards Association. Source: Certification Watch (2002).

[7] Note that certified wood is physically identical to non-certified wood, the only difference being in the environmental sensitivity of the forest management. Transgenic wood would in fact be physically slightly different involving proteins that could be detected in chemical tests.

[8] It is assumed that the benefits of transgenic technology will accrue disproportionately to the inherently high productivity sites.

References

Cashore, B. and Lawson, J. (2003) Explaining regional differences in global sustainable forestry certification rules: the case of the U.S. Northeast and the Canadian maritimes. *Canadian–American Public Policy*. The Canadian–American Center, University of Maine.

Certification Watch (2002) www.certification watch.org

DiFazio, S.P., DiFazio, S., Leonardi, S.P., Cheng, S. and Strass, S.H. (1999) Assessing potential risks of transgenic escape from fiber plantations. *Gene Flow and Agriculture: Relevance for Transgenic Crops* 72, 171–176.

Food and Agriculture Organization (2001) *The State of the World's Forests*. Forestry Department, Rome.

Pachico, D. (2003) *Regulation of Transgenic Crops: an International Comparison*. Paper delivered at the 7th International Consortium on Agricultural Biotechnology Research's Conference on Public Goods and Public Policy for Agricultural Biotechnology, June 29–July 3, Ravello, Italy. www.economia. uniroma2.it/conferenze/icabr2003/

Sedjo, R.A. (2004a) The potential economic contribution of biotechnology and forest plantations in global wood supply and forest conservation. In: Strauss, S.H. and Bradshaw, H.D. (eds) *The Bioengineered Forest: Challenges for Science and Society*. Resources for the Future, Washington, DC, pp. 22–35.

Sedjo, R.A. (2004b) *Transgenic Trees: Implementation and Outcomes of the Plant Protection Act*. Discussion Paper 04-10. Resources for the Future, Washington, DC.

Sedjo, R.A. (2004c) Wood use and trade: international trade in wood products. In: Youngs, R.L.,

Youngquist, J., Evans, J. and Burley, J. (eds) *Encyclopedia of Forest Science*. Elsevier Ltd., Amsterdam.

Sedjo, R.A. and Swallow S.K. (2002) Voluntary eco-labeling and the price premium. *Land Economics* 87, 272–284.

Williams, C.G. (2004) *Genetically Modified Pines at the Interface of Private and Public Lands: a Case Study Approach*. Paper presented at USDA Forest Service, January. Yates Building, Washington, DC.

6   Commercialized Products of Biotechnology and Trade Pattern Effects

Stuart Smyth, William A. Kerr and Kelly A. Davey
University of Saskatchewan, 51 Campus Drive, Saskatoon, Saskatchewan, Canada S7N 5A8

Introduction

Each new commercialization of a genetically modified (GM) crop creates a marketplace impact. Initially, these marketplace impacts were productivity related to the advent of commercial varieties of herbicide-tolerant oilseed rape (canola), maize and soybeans. Lately, however, the marketplace impacts have become considerably more noticeable and arise from consumer-based shifts in demand or as a result of regulatory interference. In some instances, these forces may have altered traditional international trade patterns.

The adoption of GM crops has occurred at a rapid pace in many nations. Indeed, as noted by James (2003) in his annual report on the global status of GM crops, the acreage of GM crops increased by 15% between 2002 and 2003 alone. He has estimated that the global area of GM crops reached 67.7 Mha in 2003, with 30% of this area found in developing countries. Approximately 70% of GM production occurs in North America where the USA grows 43 Mha and Canada 4.4 Mha. The US production consists of GM varieties of oilseed rape, maize, cotton and soybeans, while Canada's production includes GM oilseed rape, maize and soybeans.

The adoption of GM herbicide-tolerant soybeans leads the US market, with over 80% of soybeans planted in 2003 being genetically modified. Herbicide-tolerant cotton follows, with nearly 60% of hectares being GM varieties, while 30% of

the maize hectares were planted to Bt maize varieties (United States Department of Agriculture, 2003). This compares with adoption rates in 2000 of 55% for soybeans, 42% for cotton and 20% for maize. Not only have GM varieties been rapidly adopted, but they have been widely adopted throughout the production areas for these crop varieties.

Similar rapid adoption rates are also observed in Canada. The majority of Canadian production of GM oilseed rape occurs in the western provinces, while the majority of the production of GM maize and soybeans occurs in Ontario and Quebec. In western Canada, the adoption of herbicide-tolerant rape reached 90% (with 65% being GM varieties) in 2003, up from 77% (with 55% being GM varieties) in 2000. The total oilseed rape acreage in 2003 was 5.3 Mha, of which 3.4 million were seeded to GM rape varieties. The adoption of GM maize in central Canada rose to 32% in 2002 from 27% in 2000, while the adoption of GM soybeans rose to 32% in 2002 from 17% in 2000 (Statistics Canada, 2002). In Ontario and Quebec, 0.8 Mha of GM maize and soybean varieties were planted in 2002.

Based on these adoption rates, producer concerns about negative market impacts from the production of GM varieties would not appear to be high. Given the rapid adoption rates for most GM crop varieties, the increased level of international trade and rising concerns regarding the consumption of GM food products, the logical question to pose is: has the adoption of GM crops in Canada and the USA affected the export markets for oilseed rape, maize and soybeans? This is the question we address in this chapter.

We will compare trade patterns for Canada and the USA in oilseed rape, maize and soybeans by specifically examining the value of exports in these products. Data have been gathered from the 1990–1996 period for the USA and 1990–1996 for Canada, prior to the commercialization of GM crops. This will be compared with the post-commercialization period of 1997–2003. Once the trade patterns are identified, any changes and/or market losses can be identified. The insights gained from examining trade patterns for crops where commercial varieties of GM crops exist will be used to examine the questions relating to the introduction of GM wheat to determine what market impacts may be expected from the commercialization of this GM crop.

Background

Numerous claims have been, and still are being, made about market losses from the commercialization of GM crop varieties. Many of these claims originate from environmental groups and producer organizations opposed to the introduction and production of GM crops in both Canada and the USA. In an attempt to address the validity of these claims, we will present the arguments put forward by those opposed to GM crop technology to determine the accuracy of their statements.

One of the leading studies that was highly critical of biotechnology and suggested several areas of market loss due to the commercialization of GM crops was a report released by the UK Soils Association (2002). This report stated that the commercialization of GM oilseed rape in Canada in 1996 was directly responsible for the loss of virtually all rape sales to Europe that were reported to be C$300–400 million annually. In addition to the lost rape sales to Europe, the report goes on to suggest that Canada was on the verge of losing the ability to export rape to China due to new regulations in that country. The report suggests that the loss of this market would be approximately C$125 million per year.

The report identifies large market impacts in the USA as well, by reporting that between 1996

and 2001, US maize exports to the European Union (EU) had dropped by 99.4%. The report states that US maize exports to the EU in 1996 totalled 2.8 Mt and were worth US$305 million and, by 2001, total maize trade was 6300 t worth US$1.8 million. In addition to raw material maize exports, US maize gluten meal exports to the EU dropped from 5.5 Mt in 1995–1996 down to 4.4 Mt in 2000–2001. The report (p. 43) states that when all maize-related trade exports from the US to the EU are considered, '. . . the US lost an estimated $2 billion in trade with Europe.' There is also a suggestion in the report that US maize exports to Japan and Korea have been rejected due to contamination concerns. Finally, the report notes that the USA's share of the world soy market has dropped from 57 to 46%, but no time frame is provided for these figures. It is well known that in recent years soy production in Brazil and Argentina has been increasing rapidly, eroding the US market share.

Market loss claims have also been made by producer organizations. In March 2004, the Illinois Farm Bureau issued a press release regarding the approval of new GM maize varieties that protect against damage from root worm. The press release states that the release of these GM maize varieties '. . . could threaten a $400 million market for maize gluten. . ..' The press release additionally suggests that if the USA were to lose the EU maize gluten market, it would disrupt the entire US maize market and depress the domestic price of maize and that this would cost US maize producers US$1 billion.

Market loss claims are being made regarding the commercialization of GM wheat. In February 2004, Liberal Member of Parliament Charles Caccia stated (*Western Producer*, 2004) that 'Canada's foreign wheat customers don't want it [GM wheat] and hundreds of millions of dollars in sales are at stake . . . .'

Clearly, unsubstantiated claims are being made about market losses pertaining to the commercialization of GM crops in North America. The focus of the chapter is to examine the value of exports for crop varieties that have commercialized GM varieties and to determine what, if any, market impact has occurred. We will identify how the global market has adjusted to handle the commercialization of GM varieties and identify how international trade has moved among international buyers.

The Trade Effects of Closing Markets

Naysayers are seldom called to account when their predictions of dire consequences do not subsequently come to pass. This is particularly the case for issues pertaining to international trade (Kerr and Foregrave, 2002). There are good reasons for this. If, for example, those who believe an agreement to liberalize trade is not desirable win the argument, in part due to the negative predictions they cite, and there is no agreement, liberalization does not take place. Without liberalization, there is no way empirically to validate the predictions that made up part of the anti-liberalization argument. On the other hand, if they lose the argument and liberalization does take place, the liberalization advocates have nothing to gain by pointing out that the naysayers predictions were wrong – they have won and liberalization is taking place. The naysayers, of course, have no incentive to point out that their predictions did not come to pass.

As is the case with trade agreements, the commercialization and introduction of GM products into the market has been a subject of intense debate. The debate has been acrimonious, with vociferous advocates comprised of the (often multinational) firms engaged in the development of new crops using transgenic technology, governments in developed countries that perceive biotechnology as part of the knowledge economy upon which their industrial strategy is premised and individuals or groups that believe technology will provide solutions to problems such as world hunger, human health, and environmental stress and degradation. In opposition are ranged (often well organized) civil society groups who have concerns pertaining to: (i) the human health effects associated with consuming GM foods; (ii) the risks posed from the introduction of GM plants into the natural environment; (iii) the ethics of using science to go beyond what could occur naturally; (iv) the economic influence of large corporations over necessities such as food (Gaisford *et al.*, 2001); and (v) the loss of independence of farmers implied by the extension of strong intellectual property rights to GM seeds (Boyd *et al.*, 2003). As with any contentious issue, the debate is led by advocates and naysayers with vested interests or strong preferences who vie for the hearts and minds of a largely indifferent general public and political decision makers. In true advocate style, both sides marshal arguments and evidence that bolster their case.

As illustrated above, naysayers have not shied from using the potential 'loss of export markets' argument to further their cause.

It will be argued here that, while there may be some loss of export revenue from some markets being closed to GM products, the actual loss in revenue is likely to be far less than that prophesized by the naysayers. While the use of overly pessimistic projections of the loss of revenue for export sectors may in some cases be ingenuous, one suspects that the real cause is a poor understanding of international trade and the substitution possibilities that arise in integrated international markets.

Figure 6.1 can be used to illustrate the naïve view of the loss of revenue associated with the closure of an export market. Assume for the moment there is only one export market. In the period prior to the introduction of a GM crop and open access to the foreign market, for an exporting country the international market price, P_w, must be above the domestic market clearing price, P_e. Given price P_w, the country will produce Q_s but consume only Q_d, with the difference, Q_s minus Q_d, exported. Export revenue is area $b + c + g + f + e + j$. This observable value is the one that is typically portrayed as being at risk if a border closure takes place as the result of the introduction of a GM crop.

The reality, of course, is quite different. While it is true that there will no longer be any export revenue, the actual effect of the border closure must take into account the market adjustment that will take place as a result of the border closure. In the simplistic single foreign market case, if that market is lost, the country will be forced to move to the domestic equilibrium at P_e and Q_e. Prior to the border closure, P_e and Q_e cannot be observed – although they can be estimated. Thus, the true change in revenue is $a + b + c + g + f + e + j + h + i$ minus $e + j + h + i$ or $a + b + c + g + f$. While area $a + b + c + g + f$ can be greater or less than area $b + c + g + f + e + j$, the latter is clearly not the correct projection of change in revenue. Of course, revenue is not a particularly useful measure of the economic effects of a change in market circumstances. Economists use changes in social welfare to measure the effects of such changes – in this case the decrease in social welfare in area c, which is clearly much smaller than $b + c + g + f + e + j$.

The single export market case, however, is the 'worst case' scenario arising from a market closure. For most commodities, multiple export markets exist and the closure of one market simply means

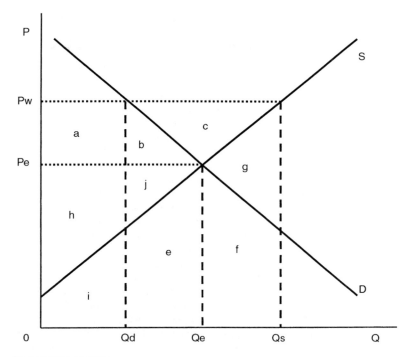

Fig. 6.1.   An exporting country.

that the cost of the change in market conditions is spread across a number of export competitors and alternative importers. This is illustrated in Fig. 6.2 which shows an exporter, country A, that is considering licensing a GM crop; an importer, country B, that will not accept GM products; and the international market. Prior to country A licensing the GM crop, total export supply in the international market equals Sxt which is comprised of the exports of country A, Sxa, plus exports of the rest of the world, Sxr. Total import demand is Dit, which is comprised of Dib, the import demand of country B plus the import demand from the rest of the world, Dir. The international market clears at Pw1 with Q2 traded, of which Q2 minus Q1 is provided by country A. Country B imports Qdb minus Qsb.

If country A licenses a GM variety of the crop and does not segregate it, then country B could consider all of country A's exports tainted and refuse to accept them. In the most extreme case, it could consider the entire international supply to be 'polluted' with GM product and totally withdraw from the market.[1] From the naysayers perspective, a value of Pw1 × (Qa2 minus Qa1) would be at risk – equal to b + c + g + f + e + j in Fig. 6.1. If this were to be the

case, country B's import demand curve would no longer exist and the international import demand curve would shift inward to Dir. The international market would move to an equilibrium at Pw2 with Q4 minus Q3 traded. Exports of country A are reduced somewhat to Qa4 minus Qa3, but part of the drop in demand is absorbed by exporters in the rest of the world, seen as a decline of Q1 minus Q3 in the international market. In country A, unlike the single export market case developed above, the price declines only to Pw2 (instead of Pe) and the decline in welfare is equal to area v (not area v + w which is analogous to area c in Fig. 6.1). In this case, the divergence between the predictions of the naysayers and the actual market result is even larger.

This multiple markets 'worst case' scenario is unlikely to come to pass, however, as the importing country may well be satisfied with simply prohibiting imports from country A (and any other adopters of GM products) and be willing to accept products from countries that have not licensed GM varieties. In this case, their demand remains in the market and traded quantities simply shift among suppliers, with country A sending to countries that will accept GM products, and countries that have not licensed

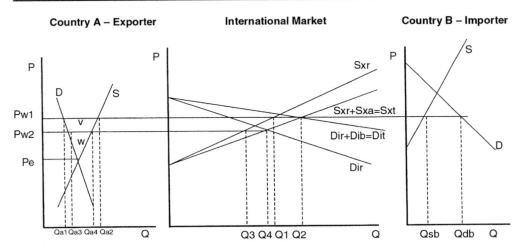

Fig. 6.2.   International market.

GM products supplying country B. In this case, the effect on price and traded quantities is likely to be minimal, with reduction in revenues simply reflecting the increased transportation and handling costs that arise from having to export to markets that can be served less efficiently.

The regulatory regime for GM products may also evolve so that segregation of GM from non-GM products may be required under the auspices of a 'consumer right to know' policy of mandatory labelling. In such a case, the heavy commercial burden will fall on the producers of non-GM crops as they will have to ensure that their products are segregated in the supply chain because consumers will care if non-GM products are tainted by GM material. On the other hand, producers of GM crops will have to trace them through the supply chain and ensure that they are labelled. They will not have to incur the heavy costs of segregation because buyers willing to consume GM products will not care if the food they are consuming is mingled with non-GM material (Kerr, 1999).

Producers of GM crops may still worry that their markets are at risk if GM crops must be labelled. Segregation means that what was a unified market will separate into two different markets. Segregated market solutions have been investigated by Gaisford *et al.* (2001). Following on from this work, Fig. 6.3 illustrates the development of a segregated market. The left hand diagram in Fig. 6.3 illustrates an importing country prior to the commercialization of a GM crop in the exporting country. Assume that the exporting

country is the sole supplier of imports. Product is supplied to the importing country at Pw, with imports equalling Qd minus Qs given the pre-GM demand curve.

Now assume that the exporting country adopts a GM variety and begins to export GM product exclusively.[2] If the importing country chooses not to license GM crops and requires that GM products be labelled if sold in their market – in effect that domestic non-GM production be segregated from GM imports – naysayers in the exporting country could claim that Qd minus Qs imports are at risk. Empirically we know, however, that only a segment of consumers will choose not to consume GM products (Gaisford *et al.*, 2001). Segregation of the market will mean that there will be separate markets for GM and non-GM products. Presumably, due to the cost advantage imparted by the technology, GM products can be supplied at a lower price than previous non-GM prices so that the import price of GM products, PwGM will be lower than the import price of the previous non-GM variety, Pw. In the non-GM market, the demand curve shifts to the left as consumers who are willing to consume GM products now begin to buy GM imports – demand curve DGM in the new GM import market.[3]

The non-GM market will reach equilibrium where domestic supply equals domestic demand[4] – at Pnon-GM and Qe. Imports equal 0 – QGM which, as drawn, is less than Qd minus Qs. Remember Pw no longer exists so that Qe could

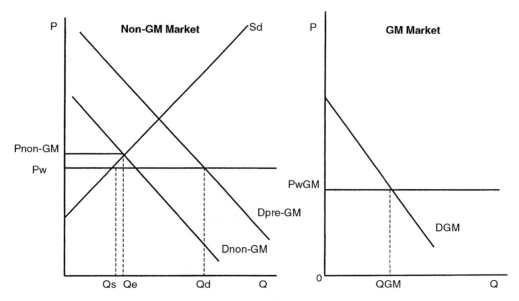

Fig. 6.3. Segregated market.

be to the left of Qs and exports could be larger than in the pre-GM era. Clearly, the effect on exports depends on the number of consumers that switch markets.

The number of consumers that are willing to switch to the GM market will depend on two factors. The first is their strength of preference for non-GM products. The second is the relative prices of GM and non-GM products, Pnon-GM minus PwGM. The greater the price differential, the more consumers will be induced to switch to the GM product. These inframarginal consumers will help to ensure that the decrease in exports is muted. Further, the larger the technical cost advantage of GM products, the more consumers are likely to switch to the GM market (Gaisford *et al.*, 2001). In the right hand diagram of Fig. 6.3, imports reach equilibrium at quantity 0 – QGM.

Thus, while in all three cases some drop in exports can be expected from the licensing of a GM product, the decreases are likely to be much smaller than those forecast by naysayers using pre-commercialization export quantities as their measure of markets at risk. Given that the effect on total exports is expected to be small although the shares of individual export markets may change considerably, the next section examines the empirical evidence from pre- and post-GM trade flows.

Analysing the Impact of GM Commercialization

While GM varieties of oilseed rape, maize and soybeans were commercialized in 1995, the adoption rates were low for 1995 and 1996. For the purposes of this chapter, we identified the start of 1997 as the first year that adoption rates were significant enough to warrant a differentiation in the marketplace. The commercialization of GM oilseed rape in Canada precipitated the development of an identity preservation system (see Smyth and Phillips, 2001 for further details on this system) that contained all GM rape produced within the North American market. In the USA, the United States Department of Agriculture (USDA) has statistics on GM maize starting for the 1997 crop year, and the adoption rate for GM soybeans in 1996 was slightly less than 10% of the total hectares planted, which was the first year data were collected. By 1997, GM soybean hectares were nearly 20% of planted acres.

In presenting these data, two assumptions were made. To present the data from Europe properly, we have assumed that the membership of the EU was comprised of 15 countries from the start of our data collection. The second assumption is that the imports of the former USSR carried over

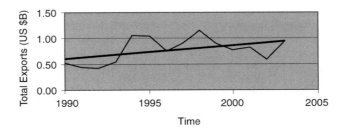

Fig. 6.4.   Canadian rape seed exports (US$ million). Source: Industry Canada (2004).

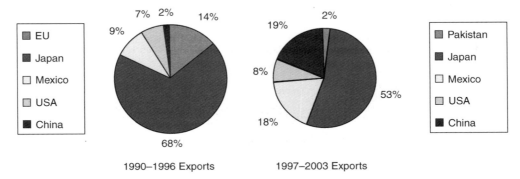

Fig. 6.5.   Canadian rape seed exports above US$80 million/year, 1990–1996 and 1997–2003. Source: Industry Canada (2004).

to be the level of imports by modern day Russia. We have not included the imports from the numerous former republics that now have their independence.

Canadian GM oilseed rape seed exports

GM oilseed rape was approved for commercialization in March 1995 and, as stated above, deliberately kept from entering the international marketplace by the use of an identity preservation system until the Japanese market approved the importation of GM rape early in 1997. The total value of Canadian rape exports has fluctuated considerably over the past 14 years, as is shown in Fig. 6.4.

The figure shows that rape exports peaked in 1998 at US$1.15 billion and then declined over the next 4 years before recovering in 2003. The shifts in key export markets for Canadian rape seed are shown in Fig. 6.5.

Figure 6.5 shows one market lost from commercializing GM rape, the EU. In the 1990–1996

period, the EU market accounted for an average of 14%, by total value of exports, of Canadian rape seed. In addition to the loss of the European market, the Japanese market dropped from 68 to 53%, resulting in a total global shift of 29%. The exports to these markets have not been lost, rather, the exports have shifted to China, Pakistan and Mexico. In the global rape seed market, the USA has remained relatively constant, while Pakistan has entered as a major market and substantial growth has occurred in exports to Mexico, from 9 to 18%, and China, from 2 to 19%.

An examination of the rape seed market, valued at less than US$80 million per year, showed numerous one time spot purchases by countries such as Norway, Switzerland and the Philippines. A total of 22 countries has made occasional purchases in very low volumes over the entire 1990–2003 time frame, some totalling less than US$100,000. While it would be confusing to present all of these countries, the more consistent purchasing countries are identified in Fig. 6.6. The number of purchasing countries prior to the commercialization of GM oilseed rape was considerably smaller than in the post-GM market. As can be seen from this figure, the EU is

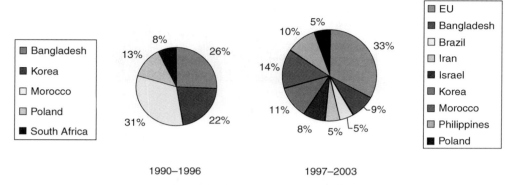

Fig. 6.6. Canadian rape seed exports below US$80 million/year, 1990–1996 and 1997–2003. Source: Industry Canada (2004).

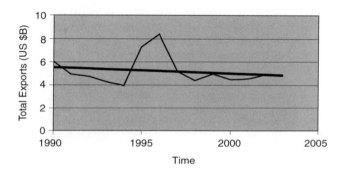

Fig. 6.7. US maize seed exports (US$ billion). Source: United States Department of Agriculture (2004).

continuing to purchase rape seed from Canada, even though claims have been made that Canada has lost all European rape export markets. Clearly, this is not the case as exports to the EU have ranged from a low of US$340,000 to US$1.1 million over the past 5 years.

The commercialization of GM oilseed rape did result in the loss of most of the EU rape seed market, but sales to the EU were successfully shifted to other markets, most notably, Mexico and China. The number of small market purchases has doubled in the post-GM market, which indicates that rape exporters have focused efforts on identifying new market opportunities. The data clearly indicate that Canada is not losing C$300–400 million annually as claimed by the UK Soils Association.

US GM maize seed exports

The initial varieties of GM maize were commercialized in 1995 and, by 1997, the USDA was separately recording planted hectares of GM maize. For this reason, we have designated the maize seed export data at the end of the 1996 calendar year as the end of the pre-GM export period, and the start of the 1997 calendar year as the start of the post-GM export period. The US maize seed export market experienced a substantial 2 year export value spike in 1995 and 1996. This sudden rise in export values came after 5 years of previous declines that saw maize export values fall from US$6.5 billion in 1989 to US$3.9 billion in 1994. While the overall trend of maize seed export values is declining, as shown in Fig. 6.7, export values have been stable in the post-GM period. In fact, maize seed export values have risen marginally starting in 2000.

The US maize seed export markets differ between the pre-GM and post-GM periods. The most noticeable changes are the absence of the EU and the former USSR in the post-GM marketplace (Fig. 6.8). Again, the value of exports to these two markets have not been lost, rather the

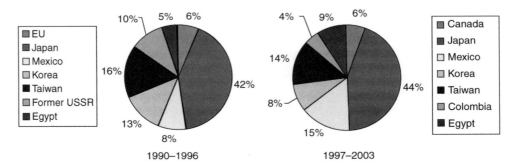

Fig. 6.8. US maize seed exports above US$1 billion/year, 1990–1996 and 1997–2003. Source: United States Department of Agriculture (2004).

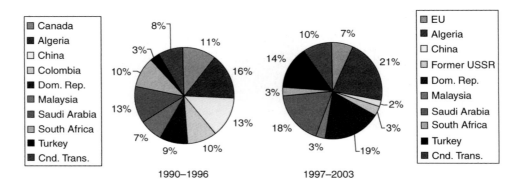

Fig. 6.9. US maize seed exports below US$1 billion/year, 1990–1996 and 1997–2003. Source: United States Department of Agriculture (2004).

marketplace has adjusted, as can be seen by the inclusion of Canada and Colombia as major US maize seed markets. The value of exports that were lost from the EU market (6%) was exactly offset by the exports to Canada (6%). The Korean market dropped by 5%, as well as small decreases (2%) in the Taiwanese and Japanese markets. The global market adjusted, as increases in the Mexican and Egyptian markets can be observed.

The number of countries represented in the smaller US maize seed markets has not changed from the pre-GM market to the post-GM market. Canada and Colombia have moved into the market valued above US$1 billion per year and have been replaced by the EU and the former USSR. Figure 6.9 identifies the marketplace for US maize seed exports valued at less than US$1 billion annually.

The US maize seed market is incredibly volatile and subject to wild fluctuations in export values over relatively short periods of time. Table 6.1 provides several examples of how rapid the fluctuations have been throughout the 1990s.

These examples show the volatility of the international maize market. Over a 3 year period, 1994–1996, both Japan and Korea increased their purchases of US maize seed by US$1 billion. The international market has managed to adjust for the commercialization of GM maize varieties, and present maize seed export values have now recovered to the same levels witnessed during the early 1990s. It is evident that the commercialization of GM maize did have some market impacts; however, the international market has adjusted and US maize exports are on the rise.

Table 6.1. Market volatility over a 3-year period in US$ million/year: actual year in parentheses.

Country	Year 1	Year 2	Year 3
Japan	1352 (1994)	1900 (1995)	2452 (1996)
China	0 (1993)	3.5 (1994)	629 (1995)
Korea	254 (1994)	1095 (1995)	1256 (1996)
Malaysia	0.37 (1993)	2.3 (1995)	187 (1996)
EU	191 (1997)	35 (1998)	1.4 (1999)
South Africa	10.3 (1997)	0.006 (1998)	31 (1999)
Former USSR	266 (1992)	407 (1993)	1.1 (1994)

Source: United States Department of Agriculture (2004).

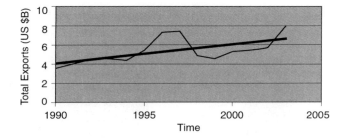

Fig. 6.10. US soybean exports (US$ billion). Source: United States Department of Agriculture (2004).

US GM soybean exports

The value of annual US soybean exports has doubled, rising from US$3.5 billion in 1990 to US$8 billion in 2003. The adoption rate of GM soybeans was nearly 20% in 1997, and this level justifies 1997 as the first year of the post-GM market. In 1996, the adoption rate was under 10% and international market impacts from this adoption level were negligible and, therefore, 1996 is the last year of the pre-GM market even though soybeans were initially commercialized in 1995. Figure 6.10 provides a yearly breakdown of export values, and the trend line for soybean exports has risen constantly over the time period.

An examination of the export markets for US soybeans shows that while the level of exports to the EU has fallen, the EU is still the largest purchaser, by value, of American soybeans (Fig. 6.11). While all of the pre-GM markets have declined in export values, these exports have shifted to China and Indonesia. Included in the post-GM market is a small percentage (4%) of soybeans that are trans-shipped through Canadian export facilities to other foreign markets, and these exports are identified as Canadian trans-shipped (Cnd. Trans.).

The market valued at less than US$1 billion annually is only changed by the three markets that moved to greater than US$1 billion annually in sales. When the soybean exports to China, Indonesia and those trans-shipped through Canada are accounted for, the remaining countries are unchanged in the pre- and post-GM periods. Figure 6.12 provides the export markets below US$1 billion in annual sales.

Market shifts are also identifiable at this level. The Canadian, Israeli and Thai markets have increased quite substantially, while the Norwegian market has virtually disappeared. Clearly, the international soy market adjusted for the commercialization of GM soybeans, while at the same time, the value of exports has continued to rise.

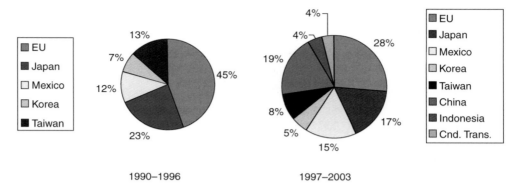

Fig. 6.11. US soybean exports above US$1 billion/year, 1990–1996 and 1997–2003. Source: United States Department of Agriculture (2004).

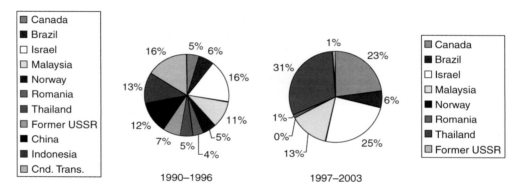

Fig. 6.12. US soybean exports below US$1 billion/year, 1990–1996 and 1997–2003. Source: United States Department of Agriculture (2004).

Export Volatility in Non-GM Commodity

Export market volatility is not restricted to commodity markets that have commercialized GM varieties. This can be demonstrated by examining the export values for Canadian and US wheat exports. To demonstrate the fluctuation in values, the same time periods have been chosen, with the split at the end of 1996 export data. Table 6.2 presents some of the market adjustments that have occurred between the two time periods.

The table identifies how Canadian wheat exports have had to shift over the past 7 years. The collapse of the former USSR, and China becoming less reliant on Canadian wheat, have resulted in Canada having to reallocate over US$6.7

billion in wheat exports. Figure 6.13 shows the annual export values and the 14 year trend.

Wheat exports have been declining over the past 14 years and for a market that was not subject to market pressures from the commercialization of any GM wheat varieties, the market is just as variable as the markets for oilseed rape, maize or soybeans. In fact, the markets that experienced GM varieties have faired considerably better than the Canadian wheat market, with oilseed rape and soybeans having a positive trend line and maize exports increasing since 2000.

A look at the US wheat export market shows similar results. The US wheat market shows comparable levels of market volatility. Table 6.3 provides some of the market variations using the same time period splits as in the previous examples.

Table 6.2.   Canadian wheat export volatility (US$ million).

Country	1990–1996	1997–2003
EU	$808	$1261
Mexico	$445	$830
Algeria	$1133	$1701
Brazil	$1074	$273
China	$4766	$612
Korea	$932	$279
Morocco	$112	$621
Former USSR	$2641	$6

Source: Industry Canada (2004).

Table 6.3.   US wheat export volatility (US$ million).

Country	1990–1996	1997–2003
EU	$647	$1459
Mexico	$847	$1808
China	$2500	$222
Morocco	$673	$269
Nigeria	$453	$1283
Pakistan	$1414	$636
Former USSR	$1777	$195

Source: United States Department of Agriculture (2004).

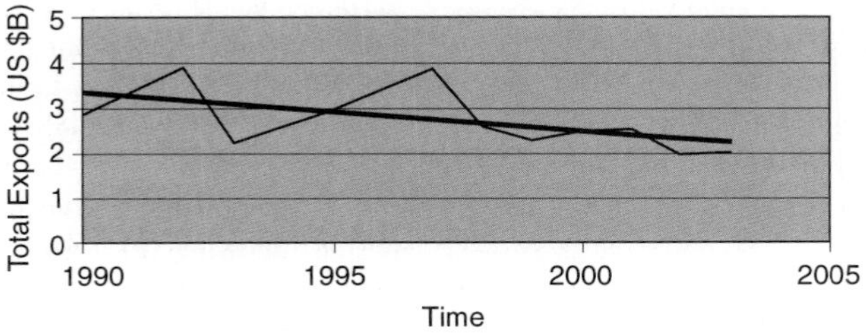

Fig. 6.13.   Canadian wheat exports, 1990–2003 (US$ billion). Source: Industry Canada (2004).

As with Canadian wheat exports, the combination of market shifts from China and the former USSR is staggering. American wheat exports have been forced to reallocate over US$3.8 billion between the two periods. As Fig. 6.14 shows, the US wheat export values have also been declining. While the situation is not as severe as that of the Canadian market, the downward value of exports is still evident.

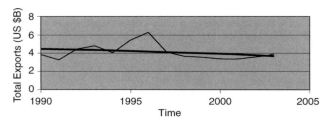

Fig. 6.14. US wheat exports, 1990–2003 (US$ billion). Source: United States Department of Agriculture (2004).

The intent of providing a comparison with a market that has no commercial GM varieties publicly released is to demonstrate that these markets have been just as, if not more, volatile as markets that have commercialized GM crop varieties. Global commodity markets are constantly and continually subject to market adjustments and these markets are fully capable of making these adjustments without unnecessary government intervention, as was being suggested when wheat markets were facing the potential of commercializing GM wheat.

Conclusions

It would appear that when GM crop varieties are initially commercialized, there is a short period where small market losses may occur; however, the international market quickly adjusts to these varieties. The oilseed rape and soybean markets have experienced export growth in the post-GM marketplace and the maize export market has been increasing since 2000. The key is that these markets have recovered in the absence of two key factors. First, there has been no regulatory intervention to restrict or attempt to control the adoption of GM rape, maize or soybeans. Secondly, all three GM crop types operate with a pooled market, i.e. any effort to provide a differentiated product such as a non-GM variety is purely voluntary.

One of the most noticeable market impacts came not from the commercialization of GM varieties, but rather from the economic collapse of the former Soviet Union. The former USSR was forced to reduce purchases in all commodity areas, which had a huge impact on the global market as exporters were quickly forced to identify market

alternatives and establish sales opportunities within those markets.

At roughly the same time as the collapse of the former USSR, the globe witnessed the rise of one of the 21st century's economic powers, China. Exports in all three GM commodities to China have risen following the commercialization of GM varieties. China has been open to the innovation of agricultural biotechnology and has had to increase market access to foreign exports following joining the World Trade Organization. Considerable amounts of commodity exports have moved from the former USSR to the Chinese market, and this impact has no relationship to the commercialization of GM crops.

There are many more factors that affect international commodity markets other than the commercialization of GM crops. While GM crops did have some market impact, it is minimal when compared with the natural volatility of these international markets. Global markets have dealt with and adjusted to market shocks considerably greater than those resulting from GM crop commercialization, and markets faced with GM variety commercialization can be assured that the international market for their exports will adjust in a similar manner.

While market losses may occur for a brief period following the commercialization of GM crop varieties, the data have shown that huge markets worth hundreds of millions if not billions of dollars are not being lost, as claimed by critics of agricultural biotechnology. The global market is capable of adjusting to introductions of GM varieties, and these shifts appear to be happening very rapidly. While critics of the industry may make public accusations, the data presented here clearly show that huge market losses are not occurring.

Notes

[1] This might be the case if there was no segregation in international facilities or transport so that ships, rail cars, canal boats or international grain terminals such as Antwerp could not be cleaned to sufficient standards between loads.

[2] It could be that the exporter will or cannot segregate in its domestic market and simply pools its GM and non-GM production – effectively meaning that the importer treats it all as GM.

[3] Note there is no domestic supply of GM products because the importer has chosen not to licence them.

[4] Remember that the exporter is not exporting any identifiable non-GM product.

References

Boyd, S.L., Kerr, W.A. and Perdikis, N. (2003) Agricultural biotechnology innovations versus intellectual property rights – are developing countries at the mercy of multinationals? *Journal of World Intellectual Property* 6, 211–232.

Gaisford, J.D., Hobbs, J.E. Kerr, W.A., Perdikis, N. and Plunkett, M.D. (2001) *The Economics of Biotechnology*. Edward Elgar Press, Cheltenham, UK.

Illinois Farm Bureau (2004) Farm group leaders against planting GM corn not yet approved in EU. Press release.

Industry Canada (2004) Data retrieved from the World Wide Web at: http://strategis.ic.gc.ca/

James, C. (2003) Preview: global status of commercialized transgenic crops: 2003. *ISAAA Briefs* No. 30. ISAAA, Ithaca, New York.

Kerr, W.A. (1999) Genetically modified organisms, consumer scepticism and trade law: implications for the organization of international supply chains. *Supply Chain Management* 4, 67–74.

Kerr, W.A. and Foregrave, R.J. (2002) The prophecies of the naysayers – assessing the vision of the protectionists in the U.S. – Canada debate on agricultural reciprocity, 1846–1854. *Estey Centre Journal of International Law and Trade Policy* 3, 357–408.

Smyth, S. and Phillips, P.W.B. (2001) Competitors co-operating: establishing a supply chain to manage genetically modified canola. *International Food and Agribusiness Management Review* 4, 51–66.

Statistics Canada (2002) Genetically modified crops: steady growth in Ontario and Quebec. Retrieved from the World Wide Web at: www.statcan.ca/english/freepub/21-004-XIE02112.pdf

UK Soils Association (2002) Seeds of doubt: North American farmers' experiences of GM crops. Retrieved from the World Wide Web at: www.soilassociation.org

United States Department of Agriculture (2003) Adoption of genetically engineered crops in the U.S. Retrieved from the World Wide Web at: www.ers.usda.gov/data/BiotechCrops

United States Department of Agriculture (2004) Foreign agricultural service. Retrieved from the World Wide Web at: www.fas.usda.gov/ustrade

Western Producer (2004) Liberal MP pushes for GM wheat rejection. February 26. Retrieved from the World Wide Web at: www.producer.com/articles/20040226/news/20040226news26.html

7 The Coexistence of GM and Non-GM Arable Crops in the EU: Economic and Market Considerations

Graham Brookes
Brookes West, Canterbury, UK

Introduction

Since 1998, a *de facto* moratorium on the regulatory approval of new genetically modified (GM) crops in the European Union (EU) has operated, effectively stopping the commercialization of GM crop traits that might be adopted by EU farmers. The only exception to this has been the planting of GM (Bt) maize in Spain (32,000 ha in 2003), which received approval prior to the moratorium in 1998.

As new legislation designed to pave the way for lifting the moratorium has been agreed (i.e. relating to labelling and traceability applicable from April 2004), one of the main subjects of current debate remains the economic and market implications of GM and non-GM crops being grown in close proximity (i.e. coexisting).

Within the coexistence debate in Europe, anti-GM groups often claim that there is no demand for GM crops in Europe and that GM and organic crops cannot coexist successfully without causing significant economic harm/losses to organic growers. This chapter examines these claims by specifically exploring the issues of demand for non-GM crops and derivatives, and identifying the context of organic arable crop production.

What is Coexistence?

Coexistence as an issue relates to 'the economic consequences of adventitious presence of material from one crop in another and the principle that farmers should be able to cultivate freely the agricultural crops they choose, be it GM crops, conventional or organic crops' (European Commission, 2003a,b). The issue is, therefore, not about product/crop safety (the GM crop having obtained full regulatory approval) but about the economic impact of the production and marketing of crops which are considered safe for the consumer and the environment.

The adventitious presence of GM crops in non-GM crops becomes an issue where consumers demand products that do not contain, or are not derived from GM crops. The initial driving force for differentiating[1] currently available crops into GM and non-GM came from consumers and interest groups who expressed a desire to avoid support for, or consumption of, GM crops and their derivatives, based on perceived uncertainties about GM crop impact on human health and the environment. This subsequently has been recognized by some in the food and feed supply chains (notably some supermarket chains and many with interests in organic farming) as an opportunity to differentiate their products and services from competitors and hence derive market advantage from the supply of non-GM products. In addition, some food companies have withdrawn from using GM-derived ingredients so as to minimize possible adverse impact on demand for their branded food products if they were to be targeted by anti-GM pressure groups.

To accommodate this perceived demand for product differentiation fully, it is important to segregate or identity preserve (IP) either GM- or non-GM-derived crops and to label these and derived (food) products throughout the food supply chain. Whilst absolute purity of the segregated product is striven for, it is a fact of any practical agricultural production system that accidental impurities can rarely be totally avoided (i.e. it is virtually impossible to ensure absolute purity).

The adventitious presence of one crop within another crop or unwanted material can arise for a variety of reasons. These include, for example, seed impurities, cross-pollination and volunteers (self-sown plants derived from seed from a previous crop), and may be linked to seed planting equipment and practices, harvesting and storage practices on-farm, transport, storage and processing post-farm gate. Recognizing this, almost all traded agricultural commodities accept some degree of adventitious presence in supplies and hence have thresholds set for the presence of unwanted material. For example, in most cereals, the maximum threshold for the presence of unwanted material (e.g. plant material, weeds, dirt, stones or seeds of other crop species) is 2%, although in durum wheat, the presence of non-durum wheat material is permitted up to 5%.

What is the Real, Current Level of Demand for Non-GM Products in the EU?

A distinct non-GM market began to develop in the EU in 1998 (for ingredients used in human food) and was extended to the animal feed sector from about 2000.[2] It focused largely on soybeans and derivatives, and to a much lesser extent maize, because these were the first two crops to receive import and use authorizations in the EU (before the introduction of the *de facto* moratorium). Key features of the soybean market development have been the following:

- In the human food sector, a switch to using alternative non-GM-derived ingredients (e.g. the replacement of soy oil with sunflower or rapeseed oil). This was relatively easy for a number of food products such as confectionery and ready meals where soy ingredient incorporation levels were low (e.g. 1%).

This course of action has been more difficult to take in the animal feed sector because of the importance of soymeal as an ingredient in some feeds (e.g. broiler feeds where typical incorporation rates are 20–25%).

- If the GM crop or derivative could not be readily replaced, non-GM-derived sources of supply were sought. This focused mainly on Brazil (but not exclusively) and involved the initiation of IP or segregated supply lines (traditional supply lines use commodity-based systems where there is broad mixing of seed in bulk for transportation) to ensure that non-GM-derived supplies to customer-specific tolerances were adhered to.

- GM-derived crop ingredients have largely been removed from most products directly consumed (by humans). However, there are two major exceptions to this: soy oil derived from GM soybeans (Table 7.1) and ingredients derived from GM microorganisms (which continue to be widely used). In the animal feed sector, the demand for non-GM soymeal affects about 25% of the EU market. In the industrial user sectors, there is little or no development of the non-GM market[3] (i.e. the market is indifferent to the production origin of raw materials).

- It has been reasonably easy for the European buyers to identify and obtain supplies of non-GM-derived soybeans and soymeal at 'competitive prices'. Where the adventitious presence threshold applied has been 1%[4] (for the presence of GM material), price differentials have tended to be in the range of 2–5% (i.e. non-GM soy has traded at a higher price than GM soy) over the last 2 years. When tighter thresholds and a more strict regime of testing, traceability and guarantees are required (e.g. to a threshold of 0.1%), the price differential has been within a range of 7–10%.

- The additional cost burden of supplying non-GM ingredients has largely been absorbed by the supply chain up to the point of retailers (i.e. the cost burden has fallen on feed compounders, livestock producers and food manufacturers and has not been passed on to retailers and end consumers).

- Any price differential that has arisen has been mainly post-farm gate. At the farm level in countries where GM crops are widely grown, there has been and is currently very little development of a price differential.

Table 7.1. Estimated GM versus non-GM soybean and derivative use 2002–2003 in the EU (t).

Product	Market size	Non-GM share	Non-GM share (%)
Whole beans	1,500,000	330,000	22
Of which used in human food	200,000	200,000	100
Of which used in feed	1,300,000	130,000	10
Oil	2,120,000	830,000	39
Of which used in food products	1,720,000	805,000	47
Of which used in feed and industrial products	400,000	25,000	6
Meal	30,770,000	8,300,000	27
Of which used in human food	800,000	800,000	100
Of which used in animal feed	29,970,000	7,500,000	25

Sources: PG Economics, Oil World, American Soybean Association.

In Brazil (the focus of non-GM supplies of soybeans), there has, to date, been no evidence of a non-GM price differential having developed. In the USA and Canada, the farm level price for non-GM supplies has tended to be within the range of 1–3% higher than GM supplies, and this level of differential in favour of non-GM crops has had little positive effect on the supply of non-GM crops (i.e. GM plantings have continued to increase, with the price differential being widely perceived to be an inadequate incentive for most farmers to grow non-GM crops such as soybeans).[5] In Brazil, trade sources[6] also suggest that a farm level differential of 5–10% for non-GM soybeans will be required to keep a significant volume of Brazilian soybean farmers growing non-GM soybeans once GM soybeans are permanently approved for planting in Brazil.[7]

Developments relating to the GM versus non-GM maize market have followed a similar path to the developments discussed above in relation to soybeans.

- The food industry targeted removal of all GM-derived ingredients from products, including GM maize.
- Non-GM-derived sources of supply were sought. This was relatively easy and focused on domestic EU origin sourcing, where the approval and commercial adoption of Bt maize has been very limited. The need to initiate IP

supply lines has also been limited because of the absence of GM maize material in the vast majority of EU supplies. Only in Spain where 20,000–25,000 ha of Bt maize have been grown annually in the period 1998–2002 has a (potential) need for greater attention to segregation/IP been relevant and, even here, there have been limited problems. The majority of Bt maize grown in Spain is concentrated in a few regions and is supplied to the local animal feed compounding sector, where there is little demand for non-GM ingredients.

- The demand for non-GM material is mostly found in the food sector (including starch). However, these uses account for a minority of total EU maize use (about 23%, Table 7.2), with the feed sector being the primary user of maize (75% of total use[8]). Overall, about 36% of total demand for maize in the EU is required to be non-GM.
- As non-GM maize accounts for 96–98% of EU maize supplies,[9] the development of a clear GM- and non-GM-derived maize market has been less marked than in the market for soybeans and derivatives. Where users of maize (notably in the food and starch sectors) have specifically required guaranteed non-GM maize (to the same thresholds as non-GM soy of mostly 1% and some to 0.1%), price differentials have tended to be in the range of 1–3% (i.e. non-GM maize prices have been higher than GM maize prices). These price

Table 7.2. Estimated GM versus non-GM maize use 2002–2003 in the EU (Mt).

Product	Market size	Non-GM share	Non-GM share (%)
Food and starch	8.97	6.28	70
Feed	29.25	7.31	25
Seed	0.78	0.55	70
Total	39	14.14	36

Source: PG Economics.

differentials have been post-farm gate, with no apparent price differential at the farm level.

- The cost burden (where applicable) of using non-GM-derived maize has generally been absorbed by the food chain.

Overall, the analysis above suggests that current EU requirements for non-GM ingredients of maize and soybeans (i.e. where buyers actively request that supplies are certified as being non-GM) accounts for about 27% of total soybean/derivative use and 36% of total maize use. The implementation of the new EU regulation on traceability and labelling of GM products is also unlikely to have any significant impact on the markets for GM versus non-GM soy and maize. This is because the vast majority of organizations that require certified non-GM products have already taken steps to locate and procure their (non-GM) requirements.[10] Only at the margin is it likely for there to be some additional demand for non-GM soy and maize products, with the net effect likely to be a fairly small increase in the size of the non-GM sector of the soy and maize markets.

In respect of other arable crops such as oilseed rape and sugarbeet, there is no real GM versus non-GM market in the EU because, in the case of oilseed rape, no GM product is currently permitted for planting or importing for use in the EU,[11] and, in the case of sugarbeet, no GM sugarbeet is currently grown commercially anywhere in the world.

Context of Organic Arable Crop Production in the EU

General

The total EU (15 countries) area devoted to organic agriculture in 2002 was estimated at about 4.3 Mha.[12] This is equal to about 3.5% of the total EU utilized agricultural area. The vast majority of the organic area (~90%) is grassland, with crops accounting for the balance.

Although this total EU area has been rising for several years (e.g. from 2.28 Mha in 1998), the total organic area in some member states has stabilized since 2000 (e.g. Denmark, Finland and Austria (source: International Federation of Organic Agricultural Movements)).

Oilseed rape

In 2003, the EU 15 planted about 3.13 Mha to oilseed rape. France, Germany, the UK and Denmark accounted for the vast majority of these plantings (94%). Within these four countries, the area planted to organic crops totalled 4811 ha, equal to 0.16% of total oilseed rape plantings (Table 7.3). The largest area of organic oilseed rape was found in Germany (3200 ha), where the organic share was 0.25% of total plantings. The country with the largest share of total organic oilseed rape plantings was Denmark (865 ha or 0.82% of total Danish oilseed rape plantings).

In terms of the accession countries, Poland has the largest area devoted to oilseed rape (360,000 ha in 2003). There are no official statistics available on the certified organic oilseed rape area in Poland, as available statistics are not disaggregated to the individual crop level. Trade sources indicate that there is no certified organic oilseed rape crop in Poland. For the other CEECs with significant oilseed rape plantings, the organic share was 1.23 and 0.19%, respectively in Slovakia and Hungary (Table 7.3).[13]

Table 7.3. EU-certified organic oilseed rape areas: main countries of production (ha).

Country	Total oilseed rape area 2003	Organic area	Organic as % of total area
France	1,083,000	496[a]	0.05
Germany	1,280,000	3,200[b]	0.25
UK	477,000	250[c]	0.05
Denmark	106,000	865[a]	0.82
Total leading four	2,946,000	4,811	0.16
EU 15	3,131,000	Not available	
Poland	360,000	Nil	Nil
Slovakia	98,000	1,207[a]	1.23
Hungary	140,000	260[a]	0.19

Sources: Coceral, ZMP, Cetiom, Soil Association, Danish Agricultural Advisory Service, Hungarian Ministry of Agriculture, Central Agricultural Control & Testing Institute Bratislava.
[a]2002; [b]2001; [c]2003.

Table 7.4. EU-certified organic maize (including forage) areas: main countries of production (ha).

Country	Total maize area 2003	Organic area	Organic as % of total area
France	3,051,000	9,998[a]	0.33
Italy[d]	1,835,000	14,994[b]	0.82
Germany	1,673,000	12,200[b]	0.73
Spain	829,000	1,000[c]	0.12
Austria	265,000	5,108[a]	1.93
Total leading five	7,653,000	43,300	0.57
EU 15	8,684,000	Not available	
Hungary	1,143,000	2,238[b]	0.2
Slovakia	132,000	1,525[b]	1.16

Sources: Coceral, ZMP, FNIP, Belgian Agricultural Economics Institute, Austrian Agricultural Economics Institute, CFRI, Hungarian Ministry of Agriculture, Italian Ministry of Agriculture, Central Agricultural Control & Testing Institute Bratislava.
[a]2002; [b]2001; [c]2003.
[d]The organic area in Italy includes crops in conversion, which accounted for about 50% of the total area. Trade sources also indicate that the 2003 organic area fell to under 10,000 ha due to difficulties in growing the crop in 2002 and rejection of some supplies by buyers because of unacceptably high levels of mycotoxins.

Maize

The total area planted to maize (grain and forage) in the EU 15 was 8.68 Mha in 2003. The main producing countries are France, Italy, Germany, Spain and Austria, which together account for 88% of total EU 15 plantings. Within these five countries, the area planted to organic crops totalled 43,300 ha, equal to 0.57% of total maize plantings (Table 7.4). The organic share across the

Table 7.5. EU-certified organic sugarbeet areas: main countries of production (ha).

Country	Total sugarbeet area 2003	Organic area	Organic as % of total area
France	367,000	Nil[a]	0.00
Germany	435,000	400[b]	0.09
UK	162,000	500[c]	0.3
Italy	205,000	0[b]	0
The Netherlands	107,000	650[c]	0.61
Spain	100,000	Nil[b]	0.00
Total leading six	1,376.000	1,550	0.11
Austria	43,000	Nil[a]	0.00
Denmark	54,000	139[a]	0.26
EU 15	1,730,000	Not available	0.077
Poland	300,000	Nil[b]	Nil
Hungary	56,000	1[b]	Nil
Slovakia	30,000	277[b]	0.9

Sources: Coceral, ZMP, BIES, British Sugar, Suker Unie, Danish Agricultural Advisory Service, Austrian Agricultural Economics Institute, Hungarian Ministry of Agriculture, Italian Ministry of Agriculture, Central Agricultural Control & Testing Institute Bratislava.
[a]2002; [b]2001; [c]2003.

leading maize-growing countries was within a range of 0.12% in Spain and Belgium, rising to 1.9% in Austria.

In the acceding countries, the largest maize producer is Hungary, where 1.14 Mha were planted to maize in 2003. Within this, 0.2% was certified as organic. In Slovakia, the organic area was 1525 ha (1.16% of total plantings). There are no official statistics available on the certified maize areas in Poland and the Czech Republic, as available statistics are not disaggregated to the individual crop level.

Sugarbeet

In 2003, the total area planted to sugarbeet in the EU 15 was 1.73 Mha. The main producing countries are Germany, France, Italy, the UK, the Netherlands and Spain, which together account for 80% of total EU 15 plantings. Within these six countries, the area planted to organic crops totalled 1550 ha, equal to 0.11% of total sugarbeet plantings in these countries (Table 7.5). The only countries with plantings of certified organic sugarbeet were

the UK, Germany, the Netherlands and Denmark, with 500, 400, 650 and 139 ha, respectively (a range of 0.09–0.61%).

In the accession countries, Poland has the largest sugarbeet area (300,000 ha). There is no reported certified organic sugarbeet grown in Poland,[14] and the recorded areas in Slovakia and Hungary are also very small.

Reasons for the very small share of organic arable crops

As indicated above, the organic share of these three crops grown in the EU is extremely low. Tables 7.6 and 7.7 also demonstrate this in terms of the relative importance of the three crops combined. For example, in France and Germany, the organic share of the total area planted to the three crops is only 0.23 and 0.47%, respectively.

This very low level of plantings and importance reflects a number of reasons, some of which are common to all three crops, and some that are crop specific.

Table 7.6. Relative importance of organic oilseed rape, sugarbeet and maize: main EU countries (%).

Country	Share of total utilized agricultural area accounted for by oilseed rape, sugarbeet and maize	Share of total area of oilseed rape, sugarbeet and maize accounted for by organic crops
Austria	10.26	1.49
Denmark	12.03	1.34
France	15.23	0.23
Germany	19.54	0.47
Spain	3.35	0.11
UK	5.00	0.23

Sources: Various: European Commission, Coceral, ZMP, Cetiom, FNAP, Soil Association, Danish Agricultural Advisory Service, MAPYA, Austrian Federal Institute of Agricultural Economics, Hungarian Ministry of Agriculture, Italian Ministry of Agriculture, Central Agricultural Control & Testing Institute Bratislava.
For years, see Tables 7.3–7.5.

Table 7.7. Relative importance of organic agriculture and organic oilseed rape, sugarbeet and maize within organic production: main EU countries (%).

Country	Organic share of total utilized agricultural area	Share of total organic area accounted for by organic oilseed rape, sugarbeet and maize
Austria	8	1.89
Denmark	6	2.88
France	2	2.03
Germany	4	2.50
Spain	1	0.26
UK	5	0.23

Sources: Various: European Commission, Coceral, ZMP, Cetiom, FNAP, Soil Association, Danish Agricultural Advisory Service, MAPYA, Austrian Federal Institute of Agricultural Economics, Hungarian Ministry of Agriculture, Italian Ministry of Agriculture, Central Agricultural Control & Testing Institute Bratislava.
For years, see Tables 7.3–7.5.

- In all three crops, weeds are a major problem and can cause significant yield loss and downgrading of a crop. Therefore, organic growers need to use rotation, mechanical methods, hand labour and land with a low incidence of weeds to minimize weed establishment when the crop canopy is not well established. These organic practices are constrained by the availability of resources such as land and labour, and lead to increased costs (which require price premiums to maintain profitability).

- In arable crops such as organic sugarbeet, effective weed control is highly dependent on mechanical control and hand labour. It is difficult to find adequate amounts of labour willing to do hand weeding (e.g. the requirement in organic sugarbeet is an estimated 60 h/ha) for short periods in the spring. Hand labour requirement also adds considerably to total weed control costs. For example, in the UK (2002), average expenditure on hand weeding was €455/ha[15] compared with the average expenditure on weed control in conventional sugarbeet crops (based largely on herbicides) of €108/ha.

- Soil nutrients, notably nitrogen, are a key factor impacting on yield in all crops. In the case of organic oilseed rape, it is not grown as readily as organic wheat because it demands high levels of soil nitrogen which are limited in an organic rotation. In conventional arable production, oilseed rape is usually grown as a break crop in rotation with wheat and allows farmers to maximize the yield potential of first year wheat.
- Levels of production risk tend to be higher in organic arable crops than conventional crops. This acts as a disincentive to convert to organic production for many combinable crops.
- There has been a general lack of demand for these organic crops. With a lack of demand, processors see little economic incentive to provide dedicated processing facilities for crops such as sugarbeet and oilseed rape, and plant breeders see little incentive to invest in the supply of organic seeds of these crops. The lack of processing facilities and the availability of some organic inputs are sometimes cited as contributory factors for the limited development of these markets and hence possible signs of market failure in the organic sector. However, it is unlikely that market failure has occurred because the small scale of demand has provided the appropriate economic signals to the supply chain and resulted in very little development of production and processing.[16]

In the oilseed rape sector, the market for organic rapeseed oil is very small. A significant proportion of the rapeseed oil used is in the non-food sector where there is virtually no organic market for any vegetable oil. Also, in the human food sector, rapeseed oil is widely considered by consumers to be an inferior product relative to alternatives such as sunflower oil (even though its health profile may be superior). The high degree of substitution between different vegetable oils used as food ingredients also means that the lowest cost organic oils dominate market use and contribute to limiting the level of organic premiums obtainable. Similarly, organic sugarbeet faces

competition from organic cane sugar which can be produced much more cheaply than organic beet (and is attractive in the high priced organic sugar market, even after payment of import duties) and is preferred by most refiners and food users of organic sugar.

It is also important to highlight that this very small development in the organic area planted to these three crops has occurred even though most member states have provided financial support schemes to assist farmers to convert and maintain organic production systems for a number of years.

The Future Level of Demand for Non-GM Products and Context of Organic Arable Crops

Future demand for non-GM-derived products

As indicated above, the existence of real markets and demand for non-GM products is limited to a minority of uses in the soybean and maize sectors.

However, anti-GM groups also often claim that there is generally little or no demand for GM products in the EU (i.e. that there is stronger demand for non-GM products). This perception does, however, fail to take into consideration several factors that suggest otherwise, including the following:

- In relation to soybeans and maize, usage is mostly concentrated in the animal feed sector and/or industrial sectors. In these markets, most users have not required their raw materials to be certified as non-GM and hence the level of positive demand for non-GM crops and derivatives has been limited. In the soybean and derivative markets, where the market for non-GM products is widely perceived to be the most developed, demand for non-GM material accounts for 27% of total consumption across the EU (see above) and is found mostly where ingredients are used directly in human food and as feed ingredients in the poultry sector. In other words, a significant majority of total consumption does not require certified non-GM material.
- Where markets have actively required the use of non-GM crops and their derivatives to

be used, these have, to date, been relatively easily obtained at prices that are similar to, or trade at only a small positive differential relative to their GM alternative. Any additional cost associated with this supply (relative to a cheaper GM-derived alternative) has largely been absorbed by the supply chain upstream of retailers, with no impact on consumer prices. When the supply chain has been able to demonstrate difficulty in absorbing even small additional costs involved in using only non-GM ingredients (e.g. in some of the livestock product sectors) to their customers in the retail sector, the non-GM requirement has tended to be dropped or made less demanding (e.g. applying only to premium ranges of products instead of all produce, such as free-range eggs or outdoor-reared pork and bacon) rather than the additional cost being accepted by retail chains and/or passed on to final consumers. This behaviour suggests that the level of demand amongst end consumers for non-GM products is highly price sensitive and would fall substantially if a consumer price level differential were to develop between GM- and non-GM derived products.

- In some markets, GM crops trade at a price premium relative to conventionally produced crops. Examples include GM soybeans in Romania and GM oilseed rape in Canada, where reduced levels of impurities in the oilseeds arriving at crushing plants have resulted in quality premiums being paid to the supplying farmers of anywhere between +1 and +3%.[17] Also, in some markets, notably China, consumer market research suggests a willingness amongst consumers to pay higher prices for GM crops because of the perceived benefits of the technology (primarily the reduction in pesticide use).[18]
- Whilst many consumer market research studies (e.g. the GM Nation Debate in the UK) suggest widespread opposition to GM products by consumers, such studies should be placed in context. Often such research uses biased language in questions, there is a poor level of understanding of the subject by respondents and actual buying behaviour is not explored to verify views expressed. In addition, some research, such as the GM Nation in the UK, is based on a biased,

self-selected audience. As such, this type of research is of limited value in identifying underlying consumer views, attitudes and actual purchasing behaviour. Where more carefully controlled research is conducted with representative samples of consumers (e.g. Institute of Grocery Distribution (IGD) in the UK in 2003), such research suggests that for a significant majority of people, the issue of whether their food is derived from GM crops is not important. For example, the IGD research found that 74% of respondents 'are not sufficiently concerned about GM food to actively look to avoid it' and it is not seen as a priority.

For the three main crops for which GM traits are seeking regulatory approval[19] for plantings in the EU in the next few years (maize, oilseed rape and sugarbeet), the level of future non-GM demand is likely to vary by crop, market and use.

- Non-GM demand will probably be highest where the crops are going into human food. Sugarbeet is probably the crop most affected here, especially as in most EU states there is a monopoly buyer of sugarbeet that can effectively dictate what varieties are planted by growers. While EU sugarbeet processors maintain a policy of not accepting GM sugarbeet (the current stated policy of most processors), there will be no market for GM sugarbeet in the EU. If this policy changes by the time of commercialization (e.g. for use in non-food sectors such as bio-ethanol) and/ or export opportunities in the bio-ethanol market arise, a GM market may develop.
- In contrast, a significant part of the animal feed and industrial sectors (about three-quarters of the ingredients used in EU animal feeds) are largely indifferent as to whether crops used are derived from GM crops or not. For crops destined for these markets, the level of active demand for crops/derivatives that have certified non-GM status is likely to remain limited. For maize, 75% of grain maize is used in the feed sector and 100% of forage maize is fed to animals. For oilseed rape, about 95% of rape meal is used in the feed sector and about 50% of rape oil is for industrial uses (e.g. biodiesel).
- The nature of competition also affects the demand for non-GM crops. In markets where

(low) price is considered to be the primary driver of demand (this is relevant to both domestically consumed foods and to export markets), access to the lowest priced products and raw materials is the main criterion used for purchasing. In such markets (e.g. frozen rather than fresh poultry), GM-based feed ingredients tend to be attractive because they are often cheaper to produce than the non-GM alternative, and hence the demand for non-GM alternatives is small.

Overall, this points to the level of demand for crops and derivatives, for which the non-GM status is important, being limited and found mostly in the sugar sector. Even in this latter sector, pressure to adopt cost-reducing technology, such as GM herbicide-tolerant sugarbeet, is likely to rise by 2008–2009, because of likely reforms to the EU sugar support system (probable significant cuts in support prices), increased competition from low priced imports (sugar from the least developed countries can enter the EU market duty-free from 2008–2009) and the further development of the market for bio-fuels, in line with EU targets for adoption of these fuels.

Future context of organic production

The certified organic production area in the EU of the main crops for which GM traits are most likely to be commercialized in the next few years is currently very low (just under 50,000 ha or 0.41% of the combined total area of the three crops of oilseed rape, sugarbeet and maize in the main EU countries growing these crops).

In the future, it is possible that the organic area of these crops could expand, although, as indicated earlier, there are a number of constraints to this.

- Crops such as oilseed rape tend to be of limited interest to organic farmers because of the crop's high nitrogen requirement relative to other break crops, and the market for organic oilseed rape is very small (those demanding organic oils prefer alternatives such as sunflower).
- For sugarbeet and cereals, which are largely processed before consumption, the EU sector is often faced with intense competition from

imported sources of (raw material) supply which tend to be more competitively priced (e.g. underlying competitive advantages of producing organic sugarcane relative to organic sugarbeet, or organic wheat produced in countries such as Argentina relative to the EU). Access to lower cost and more readily available sources of labour also contributes to competitive advantages in many developing countries.

- An important part of demand for combinable crops also comes from the livestock sector. Here the development of demand for organic produce has not matched growth experienced in the fruit and vegetable sector and is showing signs of having peaked (e.g. up to 40% of organic milk in the UK recently has had to be sold into the conventional market without an organic price premium (Wise and Findlay, 2003)) and similar organic surpluses have been reported in the organic dairy markets in Austria, Denmark and Germany.

This suggests that any further expansion in the EU organic area will be concentrated in higher value products that have characteristics such as being bulky (raises cost of transport and hence reduces the competitiveness of imports, e.g. potatoes), perishable and more commonly consumed without processing (e.g. fruit and vegetables). Even if it was assumed that there was a substantial (e.g. tenfold) increase in the EU organic area planted to combinable crops in the next 5–10 years, the sector would remain very small relative to total arable crop production.[20] It is also important to recognize that in the sectors where the organic share is higher (notably fruit and vegetables), no GM agronomic traits applicable to fruit and vegetables grown in the EU are 'on the horizon' for at least 10 years.

Coexistence of GM with Non-GM and Organic Crops to Date

The EU

For a crop to be marketed as organic, it must have been cultivated on land that has been through a period of conversion (typically 2 years) and grown according to organic principles such as only using selected (natural) pesticides and fertilizers from farm manure or nutrient-enhancing crops. However, these organic principles do not restrict the use of crop varieties or species developed by methods such as 'alien gene' transfer (e.g. used to breed yellow rust resistance and bread-making qualities into wheat from unrelated species, or the cultivation of triticale, a man-made hybrid of wheat and rye[21]).

Baseline organic requirements are set at an EU level, although each organic certification body has the freedom to set its own principles and conditions that may be stricter than the legal baseline. As a result, there may be several different organic standards operating in member states, each striving for market differentiation relative to others.

In relation to the adventitious presence of genetically modified organisms (GMOs), the base EU regulation covering organic agriculture (2092/91) states 'there is no place for GMOs in organic agriculture' and that '(organic) products are produced without the use of GMOs and/or any products derived from such organisms'. The legislation made provision for a *de minimis* threshold for unavoidable presence of GMOs which should not be exceeded, but did not set such a threshold. In the absence of such a legal threshold having been set, the general threshold of 0.9%, laid down in the 2003 Regulation on labelling and traceability, is the current legally enforceable threshold. Although the current legally enforceable threshold for GMO presence labelling is 0.9%, some organic certification bodies apply a more stringent *de minimis* threshold on their members (0.1%, the limit of reliable detection).

For conventional growers of non-GM crops in the EU, the 'benchmark' for determining whether a crop or derivative has to be labelled as GM or not is also the legally enforceable threshold of 0.9%, although some buyers may choose to set more stringent thresholds (e.g. 0.1%).

Against this background, evidence from the only current example of where GM crops are grown commercially in the EU (Bt maize in Spain[22]) shows that GM, conventional (non-GM) and organic maize production have coexisted without economic and commercial problems. This includes regions such as Catalunya where Bt is concentrated.[23] Where non-GM maize has been required in some markets, supplies have been relatively easily obtained, based on market-driven adherence to on-/post-farm segregation and by the purchase of maize from regions where there has been limited adoption of Bt maize (because the target pest of the Bt technology, the corn borer, is not a significant problem for farmers in these regions). Only isolated instances (two) of GMO adventitious presence in organic maize crops were reported in 2001.

North America

As in the EU, National Organic Standards (e.g. in the USA) prohibit the use of GM varieties. However, an important point to note in the US regulations is the recognition that organic growers may need to implement practical procedures to minimize the possibility of the adventitious presence of GMOs in their crops occurring and that if an organic crop tests positive for a GM event that occurs unintentionally, the grower should not be penalized by the downgrading of a crop (i.e. loss of an organic price premium) and/or by the decertification of a specific field. For growers of conventional, non-GM crops, there is no formal regulatory 'benchmark' for the definition of whether a product should be labelled as GM or not, except where crops/derivatives are exported to countries where labelling legislation does exist (e.g. the EU) or buyers set purchasing criteria on commercial grounds.

Relative to this more limited (relative to the EU) regulatory background and interpretation, GM crops have been grown commercially in North America since 1996 and now account for 60% of the total plantings of soybeans, maize and oilseed rape in the USA and Canada combined. Against this background, GM crops have coexisted with conventional and organic crops without causing significant economic or commercial problems.[24,25] For example, the US organic areas of soybeans and maize increased by 270 and 187%, respectively, between 1995 and 2001, a period in which GM crops were introduced and reached 68 and 26% shares of total plantings of soybeans and maize. Also, survey evidence amongst US organic farmers shows that the vast majority (92%) have not incurred any direct, additional costs or incurred losses due to GM crops having been grown near their crops.

Can the EU Organic Sector Coexist with Future GM Production?

The evidence to date shows that GM arable crops growing commercially in the EU and in North America have coexisted with conventional and organic crops without economic and commercial problems – only isolated instances have been found of the adventitious presence of GMOs occurring in organic crops in Spain and a small number found in North America, even though GM crops dominate production of soybeans, maize and oilseed rape in North America. Furthermore, in a number of cases, these instances have been attributed to weaknesses in on- and post-farm segregation of crops or to failure of organic growers to use organic seed or to test their conventional seed for GMO presence prior to sowing.

For the future, the likelihood of economic and commercial problems of coexistence arising remains very limited, even if a significant development of commercial GM crops (see Appendix 7.1 for a summary of the likely timing of different GM traits being commercialized in the EU over the next few years) and increased plantings of organic crops were to occur.

- The GM traits being commercialized in the next few years are in crops for which there is limited demand for non-GM material (e.g. for forage and grain maize, rapeseed oil and meal). The only possible exception to this is sugarbeet, although, even here, the development of non-food uses of sugar (e.g. for bio-ethanol) and policy change-induced competitive pressures may result in greater willingness amongst the EU's sugar processors to use GM sugarbeet.
- The organic areas of the three key crops (oilseed rape, sugarbeet and maize) are extremely small (only 0.41% of the area planted to these crops in the main producing countries of the EU).
- The organic area of these crops (and other combinable crops) is likely to continue to be a very small part of the total arable crop areas (even if there were a tenfold increase in plantings), with a very limited economic contribution relative to the rest of the EU's arable crops. The likelihood of these (organic) areas expanding is limited due to a combination of adverse agronomic factors (e.g. a need

for sites with few weed problems and the nutrient-demanding nature of crops such as oilseed rape), limited demand, and market preference for competing (imported) produce (e.g. cane sugar).

- The possibility of gene transfer to related wild and other crop species from any of the GM crops is extremely low;[26] this is also an issue examined before regulatory approval is given.
- EU arable farmers have been successfully growing specialist crops (e.g. seed production, high erucic acid oilseed rape, waxy maize) for many years, near to other crops of the same species, without compromising the high purity levels required.
- Some changes to farming practices on some farms may be required once GM crops are commercialized. This will, however, only apply where GM crops are located near non-GM or organic crops for which the non-GM status of the crop is important (e.g. where buyers do not wish to label products as being GM or derived from GM according to EU labelling regulations). These changes are likely to focus on the use of separation distances and buffer crops (of non-GM crops) between the GM crops and the 'vulnerable' non-GM/organic crop and the application of good husbandry (weed control) practices.
- GM crop-planting farmers are already made aware of these practices as part of recommendations for growing GM maize in Spain (coexistence and refuge requirements) provided by seed suppliers in their 'GM crop stewardship programmes'. Few GM-planting farmers have, however, found themselves located near to 'vulnerable' non-GM/organic crops and hence the need to apply these guidelines strictly has been very limited.

The different certification bodies in the EU organic sector can also take action to facilitate coexistence by the following types of action.

- Applying a more consistent, practical, proportionate and cost-effective policy towards GMOs (i.e. adopt the same policy as it applies to the adventitious presence of other non-organic material). This would allow it to better exploit market opportunities and to minimize the risks of publicity about inconsistent organic definitions and derogations for the use of non-organic ingredients and inputs damaging consumer confidence in all organic produce. This latter point is important given that the organic crops perceived to be affected by the commercialization of GM traits in the next few years account for a very small share of the total organic farmed area in the EU (Table 7.7). For example, in Austria and Germany, the share of the total organic area accounted for by organic oilseed rape, sugarbeet and maize was 1.89 and 2.5% respectively. Applying the same testing principles and thresholds currently applied to GMOs to impurities (e.g. introduce a *de minimis* threshold on pesticide residues and apply a 0.1% threshold on the limit for acceptance of all unwanted materials and impurities).[27]

• Accepting that if they wish to retain policies towards GMOs that advocate farming practices that go beyond those recommended for GMO crop stewardship (e.g. buffer crops and separation distances that are more stringent than those considered to be reasonable to meet the EU labelling and traceability regulations), then the onus for implementation of such measures (and associated cost) should fall on the organic certification bodies and their members in the same way as current organic farmers incur costs associated with adhering to organic principles and are rewarded through the receipt of organic price premiums.

Lastly, it is important to emphasize the issues of context and proportionality. If highly onerous GM crop stewardship conditions are applied to all farms[28] that might wish to grow GM crops, even though the vast majority of such crops would not be located near to organic-equivalent crops or conventional crops for which the non-GM status is important, this would be disproportionate and inequitable. In effect, conventional farmers, who account for 99.59%[29] of the current, relevant EU arable crop farming area, could be discouraged from adopting a new technology that is likely to deliver farm level benefits (yield gains and cost savings) and provide wider environmental gains (reduced pesticide use, switches to more environmentally benign herbicides and reduced levels of greenhouse gas emissions).[30]

Notes

[1] Generally referred to either segregation or identity preservation.

[2] Brookes (2001).

[3] This refers to all non-food industrial uses and does not refer to industrial uses where the raw materials are destined for human food use (e.g. maize starch used in food products).

[4] More recently 0.9% in line with the new legal threshold.

[5] Some US farmers of GM soybeans have also reported positive price differentials in favour of GM soybeans because of the lower levels of impurities found in their crops.

[6] With interests in the supply of non-GM soybeans.

[7] In 2003, when GM soybeans were given temporary approval for planting, about 18% (3.24 Mha) of the Brazilian soybean crop is reported to have been GM.

[8] The balance is accounted for by seed.

[9] The GM share comes from Spanish production of about 0.32 Mt in 2003 and annual imports of between 0.6 and 1.4 Mt from Argentina.

[10] In many cases, this policy was also extended to products derived from GM crops such as soy oil even though they did not have to be labelled prior to 18 April 2004.

[11] It is, however, interesting to note the ultra-cautious behaviour of some crushers in the UK, where for crops supplied in 2004 (now that the new labelling and traceability law is operational), farmers were required to make declarations as to the non-GM status of their oilseed rape crops, purely because of the (remote) possibility of GM adventitious presence arising from a GM oilseed rape farm-scale trial.

[12] Source: International Federation of Organic Agricultural Movements.

[13] There are no official statistics available on the certified organic oilseed rape area in the Czech Republic, as available statistics are not disaggregated to the individual crop level.

[14] Source: Sugar processing sector.

[15] Source: Lampkin and Measures (2003).

[16] If a fundamental imbalance between supply and demand developed, a substantial organic premium would have occurred. There is no evidence that such a large price premium has developed for organic crops that require processing, suggesting that there is no market failure.

[17] Brookes (2003a,b); Brookes and Barfoot (2003a, b); Canola Council (2001).

[18] Quan (2002).

[19] Or having already gained regulatory approval in the case of some Bt, insect-resistant maize events.

[20] It should also be noted that despite the provision of subsidies to support both the conversion and maintenance of organic production systems in most EU member states for several years, even in countries such as Austria, where 8.3% of the total agricultural area was classified as organic in 2002, the share of this area accounted for by organic oilseed rape, sugarbeet and maize was still only 1.89%.

[21] Triticale is an artificial hybrid of wheat and rye and is a popular organic crop. Triticale is an example of a wide-cross hybrid, made possible solely by the existence of embryo rescue (a method of recovering embryos in laboratory culture) and chromosome doubling techniques (the restoration of fertility using mutagenic chemicals). The triticale crop could not exist without human manipulation of the breeding process, nor could wheat varieties produced using alien gene transfer techniques.

[22] See Brookes and Barfoot (2003a).

[23] Bt maize accounts for about 15% of total maize plantings in this region.

[24] See Brookes and Barfoot (2004).

[25] This relates to reports of the adventitious presence of GM material occurring in organic crops that have resulted in economic losses for organic growers (e.g. loss of organic price premium). It does not include instances where trace levels (within the boundaries of very sensitive testing equipment) of GM material may have been detected, but which did not result in any economic loss.

[26] For example, the FSEs in the UK found no evidence for the transfer of the herbicide tolerance gene from GM oilseed rape to common wild relatives.

[27] Alternatively apply the current legal labelling threshold of 0.9% for GM material.

[28] For example, the setting of substantial separation distances between GM crops and any conventionally grown equivalent.

[29] This varies by member state within a range of 99.77% in the UK to 98.07% in Austria.

[30] See Appendix 5 of PG Economics (2003) for detailed analysis of this.

References

Brookes, G. (2001) *GM Crop Market Dynamics, the Case of Soybeans*. European Federation of Biotechnology, Briefing Paper 12.

Brookes, G. (2003a) Co-existence of GM and non GM crops: economic and market perspectives. On www.pgeconomics.co.uk

Brookes, G. (2003b) *The Farm Level Impact of Herbicide Tolerant Soybeans in Romania*. Brookes West, Canterbury, UK. www.pgeconomics.co.uk

Brookes, G. and Barfoot, P. (2003a) GM and non GM crop co-existence: case study of maize grown in Spain, *1st European Conference on Co-existence of GM Crops with Conventional and Organic Crops, Denmark*. Also on www.pgeconomics.co.uk

Brookes, G. and Barfoot, P. (2003b) *GM and Non GM Crop Co-existence: Case Study of the UK*. PG Economics, Dorchester, UK. www.pgeconomics.co.uk

Brookes, G. and Barfoot, P. (2004) *Co-existence Case Study of North America: Widespread GM Cropping with Conventional and Organic Crops*. PG Economics, Dorchester, UK. www.pgeconomics.co.uk

Canola Council of Canada (2001) *An Agronomic and Economic Assessment of Transgenic Canola*. Canola Council, Canada. www.canola-council.org

European Commission (2003a) *Communication on Co-Existence of Genetically Modified, Conventional and Organic Crops*. March 2003.

European Commission (2003b) *Recommendation on Guidelines for the Development of National Strategies and Best Practices to Ensure the Co-existence of GM Crops with Conventional and Organic Agriculture*. July 2003.

International Federation of Organic Agricultural Movements (undated) Position on genetic engineering and GMOs. www.ifoam.org

Lampkin, N. and Measures, M. (2003) *2002 Organic Farm Management Handbook*, 5th edn. University of Wales, Cardiff, UK.

Quan, L. (2002) Consumer attitudes towards GM foods in Beijing, China. AgbioForum 5, No. 4, Article 3.

Wise, C. and Findlay, A. (2003) *BCPC International Congress – Crop Science & Technology*. BCPC, pp. 723–726.

Appendix 7.1: Possible GM Technology Use in the EU

Table 7.8 summarizes our forecasts for when reasonable volumes of seed containing GM traits in the leading arable crops of relevance to the EU are likely to be available to EU farmers. The key point to note is that it is likely to be another 2–3 years before GM seed is widely available to EU producers of crops such as oilseed rape and sugarbeet, and the only GM crop with current commercial availability is insect-resistant maize. GM wheat and potatoes are unlikely to be available until after 2010.

Table 7.8. Forecast GM crop commercial availability for leading agronomic traits in the UK.

Crop/trait	Commercially available to EU farmers
Herbicide (glufosinate)-tolerant maize	2005–2006
Herbicide (glufosinate)-tolerant oilseed rape	2005–2007
Novel hybrid oilseed rape	2005–2007
Herbicide (glyphosate)-tolerant sugarbeet	2006–2008
Herbicide (glyphosate)-tolerant wheat	After 2010
Fungal-tolerant wheat	After 2010
Nematode- and fungal-resistant potatoes	After 2010
Fungal-resistant oilseed rape	After 2010

Source: PG Economics.
The glufosinate-tolerant trait in maize has received regulatory approval in the EU, but seed varieties containing the trait have yet to receive varietal approval for use in any member state.

8 Research Spillovers in the Biotech Industry: the Case of Oilseed Rape

Richard S. Gray[1], Stavroula Malla[2] and Kien C. Tran[2]

[1]University of Saskatchewan, Saskatoon, Saskatchewan; [2]University of Lethbridge, Alberta, Canada

Introduction

Intellectual property rights (IPRs) have fundamentally altered the nature of agricultural research spillovers. During most of the 20th century, the public good attributes of research were recognized as a spillover, with the result that most crop research was undertaken by public institutions and the products of the research were held in the public domain (Huffman and Evenson, 1993). The ability of modern biotechnology to identify DNA, combined with regulatory and judicial moves to enhance IPRs, have reduced these spillovers, resulting in substantial private investment in agricultural research (Fernandez-Cornejo, 2004). The inherent non-rival nature of research products, along with *freedom to operate* costs, has led to rapid consolidation and a concentrated agricultural research industry (see, for example Lesser, 1998; Fulton and Giannakas, 2001), so much so that US anti-trust regulators have made recent biotech acquisitions subject to divestiture in order to limit market concentration (Schimmelpfennig et al., 2004). At the same time, more than 200 public research institutions have moved to create offices of technology transfer to manage access to their intellectual property (Graff et al., 2003) and national governments have passed laws to protect landrace genetics (Evenson, 1999; Falcon and Fowler, 2002). The combination of these effects represents a watershed of change within the agricultural research industry.

The change in crop research has been particularly evident in the Canadian oilseed rape (canola) industry. After three decades of public leadership, the rape industry has become dominated by large private firms employing biotechnology to produce tailored products for the marketplace. Since 1985, the private sector has funded about 60% of the total investment in research, and owned 85% of the new varieties (Gray et al., 2002). By the year 2000, 75% of the rape acreage was planted to varieties that required farmers to make annual purchases to retain access to the technology (Malla et al., 2003). Despite the importance of these changes in crop research, a lack of firm-specific data hampered the analysis and understanding of the biotech industry.

This study uses firm-specific data in the Canadian oilseed rape industry to examine a number of research spillovers among public and private firms. The effect of 'spill-ins' is examined at the firm level in terms of their impact on research output, sales revenue and a measure of downstream 'social' value. The potential sources of pecuniary and non-pecuniary spillovers examined include basic research by public institutions, human capital and knowledge (as measured through other firm expenditures) and genetic spillovers (as measured through variety yields of other firms). Inferences about downstream spillovers and market-driven pecuniary spillovers are made from a comparison of the estimated relationships.

© CAB International 2006. *International Trade and Policies for Genetically Modified Products* (eds R.E. Evenson and V. Santaniello)

The methodological contribution of the chapter is to develop a set of empirical models that distinguish between many types of potential spillovers within a crop research industry. Empirically, we are able to show that positive inter-firm and downstream research spillovers have been important in the modern oilseed rape research industry and that public basic and applied research has caused a 'crowding in' of private research activities.

The remainder of the chapter is organized into four sections. The first outlines the relevant literature. The second describes the theoretical framework used for this analysis. The empirical model and the results are reported in the third section, while the last section contains the concluding comments of the chapter.

Research Spillovers

Research spillovers are central to the economics of research. These externalities that arise from the public good aspects of knowledge are an important determinant of economic productivity (e.g. Jaffe, 1986; Adams, 1990; Griliches, 1992). The non-rival nature of research output has assumed a central role in endogenous growth theory, in terms of both physical capital (e.g. Romer 1986, 1990; Aghion and Howitt, 1992) and human capital (e.g. Lucas, 1988). Spillovers also have important implications for firm behaviour (e.g. Cohen and Levinthal, 1989; Just and Hueth, 1993; Moschini and Lapan, 1997; Adams, 2000), industrial organization (e.g. Dasgupta and Stiglitz, 1980; Levin and Reiss, 1984, 1988; Spence, 1984; Fulton, 1997; Lessor, 1997; Fulton and Giannakas, 2001) and industrial structure (Acs et al., 1994; Schimmelpfennig et al., 2004).

A significant body of economic research has addressed the spillovers from public research by examining the crowding effects of public research investment on private research investment. Roberts (1984), Bergstrom et al. (1986) and David and Hall (2000) argue that publicly funded research competes for scarce resources and therefore could 'crowd out' privately funded research. Other economists who have considered charitable donations (e.g. Khanna et al., 1995; Khanna and Sandler, 1996) show that public expenditure could have the opposite effect and cause a 'crowding in' of private research expenditure. David et al. (1999)

provided a recent survey of the available empirical evidence and found that the results were inconclusive in terms of the direction and magnitude of the relationship between public and private research expenditure.

The effects of spillovers on agricultural productivity have also attracted significant attention in the literature (e.g. Griliches, 1979, 1980; Evenson 1989; Huffman and Evenson, 1993; Johnson and Evenson, 1999; White et al., 2003), while a number of studies examined the cross-state spillovers from agricultural research (e.g. Evenson, 1989; Alston and Pardey, 1996; Yee and Huffman, 2001).

Some economists have distinguished between the various sources of research spillovers. Pardey et al. (1996) examined the genetic research spillovers through pedigree attribution among different breeding programmes, which applies when crop pedigrees are known (Heisey and Morris, 2002). The seminal work of Evenson and Kislev (1976) introduced the notion of basic research spillovers, using a theoretical model where the outputs of basic research (i.e. scientific knowledge) improve the productivity of the search process (i.e. applied research). This concept was used in a number of later studies (e.g. Lee 1982, 1985; Kortum, 1997). Diamond (1999) and Robson (1993) empirically examined the crowding effects of basic research. Finally, a number of studies have recognized that knowledge is embodied in human capital and that spillovers occur with the education of workers (e.g. Shultz, 1975; Lucas, 1993), learning from others (Foster and Rosenzweig, 1995; Thorton and Thompson, 2001) and with the mobility of workers (Glaeser et al.,1992).

To estimate inter-industry spillovers, it is important to distinguish between pecuniary spillovers and non-pecuniary spillovers. Pecuniary spillovers are the firm-to-firm interactions that occur through prices in a properly functioning market as the firms purchase inputs, and sell output. Non-pecuniary spillovers are the non-market impacts of a firm's actions on other firms. These spillovers typically exist because of poorly defined property rights or other forms of market failure. These spillovers are the focus of most economic analysis, because they distort private incentives away from efficient marginal conditions. The combined effects of pecuniary and non-pecuniary spillovers are important because they measure the net effect of a firm's actions on its rivals. When the combined spillover from an increase in output is negative, firms will tend to 'crowd out' their rivals. In the case of positive combined spillovers, this will tend to create a 'crowding in' of other firms, or a research clustering effect.

Modelling the Impact of Research Spillovers

In Canada's oilseed rape industry, both public and private research firms expend resources to develop enhanced rape varieties. Private firms engage in applied research to produce enhanced crop varieties, which are sold to farmers. Public institutions (firms) also engage in a significant amount of basic research, which creates knowledge that is used to improve crop research processes. The varieties created by public firms are distributed through the private seed industry in return for royalty payments.

The empirical model considers several possible pecuniary and non-pecuniary spillovers, which are illustrated in Fig. 8.1. The solid arrows in the figure represent both upstream and downstream market linkages, which are the source of pecuniary spillovers. The potential non-pecuniary spillovers are shown as dashed arrows. The non-pecuniary spillover between the firms and the downstream industry (labelled as 1) are those benefits provided to those firms and consumers that are outside of the market (e.g. benefits associated with farmers retaining seed of a new variety for the next year). The possible combinations of inter-firm non-pecuniary spillovers are labelled separately and shown as: private to private, 2a; private to public, 2b; public to private, 2c; and public to public, 2d. These inter-firm non-pecuniary spillovers can be broken down further into: (i) spillovers generated from the germplasm of other firms; (ii) spillovers generated from knowledge created by the applied research of other firms; and (iii) spillovers generated from the knowledge created by the basic research of public firms.[1]

In the econometric model following, we model the firm-to-firm spillovers in three different models: *model 1* at the level of each firm's production function for new varieties; *model 2* at the level of each firm's sales revenue; and *model 3* at the level of the social benefits embodied in each firm's varieties. Model 1 allows us to isolate the inter-firm non-pecuniary spillover effects. Model 2, based on

revenue, will capture both the non-pecuniary spillovers and the pecuniary spillovers generated by the other firms in the output market, providing an indication of the crowding effects. The measure of firm output in model 3 includes the spillovers to downstream firms and consumers, giving an indication of some of the social impacts of inter-firm spillovers. In the estimation, we make a further distinction between public and private research enterprises such that the inter-firm spillovers within the public or private sector are potentially different from the private–public spillovers.

Model 1

Research outputs, the dependent variables in model 1, are measured as the average yield of new varieties sold by each firm each year. The production function of each research firm is described by the function:

$$Y_{it} = f(Y_{i,t-1}, AR_{i,t-k}, BR_{t-l}, OAR_{i,t-r}, OY_{i,t-m})$$
$$\quad\;\; + \qquad\quad + \qquad\quad + \qquad\quad + \qquad\quad + \qquad\qquad (1)$$

where Y_{it} is the average yield of new varieties of firm i in year t, which is an increasing function of the previous period's yield, $Y_{i,t-1}$, lagged own applied research, $AR_{i,t-k}$,[2] and, through non-pecuniary spillovers, an increasing function of lagged basic research expenditures, BR_{t-m}, lagged applied expenditure of other firms, $OAR_{i,t-r}$, and the lagged yield of other firms, $OY_{i,t-m}$.

In model 1, the derivative of output with respect to the own applied research is the single-year marginal product (in terms of yield increase) of an additional unit of applied research. The derivative with respect to the lagged own yield represents the marginal persistence effect. The marginal impact of the spillovers on a firm's output, from public basic research, its rival's applied research and its rivals' yield levels, are also captured in the derivative of the function.

Model 2

The dependent variables in this model are the revenues of each firm i in each year t, which consists of the sum of sales revenue and technical use fees. The revenue generated will be a function of those

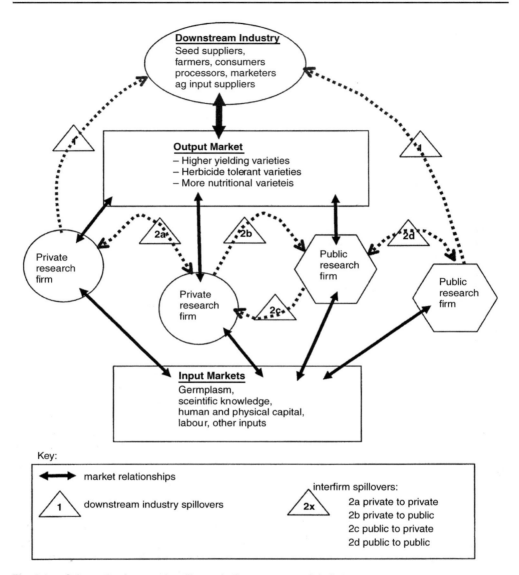

Fig. 8.1. Schematic of research spillovers in the crop research industry.

variables that affect variety yields, as well as the variables that measure IPRs and the pecuniary effects of competition. Specifically,

$$R_{i,t} = g(R_{i,t-1}, AR_{i,t-k}, PBR_t, HYB_{i,t}, TUREV_{i,t},$$
$$\quad\quad + \quad\quad + \quad\quad\quad + \quad\quad\quad +\quad\quad\quad\quad\quad +$$
$$BR_{t-1}, OAR_{i,t-r}, OY_{i,t-g}) \quad\quad (2)$$
$$\quad ? \quad\quad ? \quad\quad ?$$

In this case, sales revenue, $R_{i,t}$ is an increasing function of last year's revenue, $R_{i,t-1}$, own lagged applied research, $AR_{i,t-k}$ plant breeders' rights, PBR_t, the proportion of area seeded to varieties requiring annual repurchase, $HYB_{i,t}$, and technical use revenue, $TUREV_{i,t}$. The latter three variables measure strengthened IPRs and greater ability to capture revenue, and therefore each should be positive. The spillovers from lagged basic research, BR_{t-b} other firms' applied research expenditure, $AOR_{i,t-k}$, and yield of other firms' varieties, $OY_{i,t-g}$, could be positive (negative) if the non-pecuniary effects are greater (less) than the pecuniary spillovers. For model 2, the derivatives of the function represent the private marginal revenue effects.

Model 3

This model estimates the production of social revenue associated with the varieties sold by each firm. In this case, the annual social benefit is approximated as the increase in economic surplus generated from the yield's increases plus the herbicide savings rent. The economic surplus attributed to yield is estimated as the yield increase on the area sown to the firm's varieties multiplied by prevailing oilseed rape prices. The spillovers at this level will include both pecuniary and non-pecuniary inter-firm spillovers, as well as the downstream spillovers. The social revenue for firm i in period t can be described as:

$$SR_{i,t} = k(SR_{i,t-1}, AR_{i,t-k}, PBR_t, BR_{t-1}, OAR_{i,t-r})$$
$$\quad + \qquad + \qquad ? \qquad ? \qquad ? \qquad (3)$$

Social revenue, $SR_{i,t}$, is expected to be an increasing function of the firm's lagged applied research, $AR_{i,t-k}$, and the previous period social revenue, $SR_{i,t-1}$. The market-correcting effect of plant breeders' rights, PBR_t, is ambiguous in the presence of other spillovers, given Lancaster's argument of the second best. The spillovers from lagged basic research, BR_{t-1}, and other firms' lagged applied research, $OAR_{i,t-r}$, will be negative (positive) if the social pecuniary effects are greater (less) than non-pecuniary spillover effects. The derivatives of Equation 3 represent the marginal impacts of each variable firm on the sum of firm revenue and downstream spillovers.

Econometric Analysis

Data and econometric model specification

The data used for the econometric analysis came from many industry and government sources and in some cases took considerable calculation to construct each of the variables used for the econometric analysis. We were able to construct a data set for five private firms and two public institutions. The primary data source for research expenditures was a survey of the Canola Industry (Canola Research Survey, 1999). These data sources and the methodology used to construct each variable are described in Appendix 8.1.

We used three different models for the analysis of research spillovers. To separate research spillover effects in public firms and private firms, we divided the firms into two groups: private firms and public firms. In the first model, we considered the following specification.

Model 1

$$Y_{i,t}^{PV} = \beta_{0i} + \beta_1 AR_{i,t-k}^{PV} + \beta_2 BR_{t-l} + \beta_3 OAR_{i,t-r}^{PV}$$
$$+ \beta_4 AR_{t-m}^{PUB} + \beta_5 OY_{i,t-h}^{PV} + \beta_6 Y_{t-g}^{PUB}$$
$$+ \gamma_i Y_{i,t-1}^{PV} + u_{i,t}, \quad i = 1, \ldots, 5 \qquad (4)$$

$$Y_{j,t}^{PUB} = \delta_{0j} + \delta_1 AR_{j,t-k}^{PUB} + \delta_2 BR_{t-l} + \delta_3 OAR_{j,t-r}^{PUB}$$
$$+ \delta_4 AR_{t-m}^{PV} + \delta_5 OY_{j,t-h}^{PUB} + \delta_6 Y_{t-g}^{PV}$$
$$+ \gamma_j Y_{j,t-1}^{PUB} + u_{j,t}, \quad j = 1, 2 \qquad (5)$$

where we assume that $|\gamma_i| < 1$ and $|\gamma_j| < 1$, for all i, j to ensure stationary, and $Y_{i,t}^{PV}$ = annual weighted yield index of private firm i in year t; $Y_{j,t}^{PUB}$ = annual weighted yield index of public firm j in year t; BR_{t-l} = basic research expenditures in year $t-l$ (same for all seven firms); $AR_{i,t-k}^{PV}$ = private applied research expenditures of firm i in year $t-k$; $AR_{i,t-k}^{PUB}$ = public applied research expenditures of firm j in year $t-k$; $OAR_{i,t-r}^{PV}$ = total applied research expenditures of other private firms excluding firm i in year $t-r$; $OAR_{j,t-r}^{PUB}$ = total applied research expenditures of other public firms excluding firm j in year $t-r$; AR_{t-m}^{PV} = total applied research expenditures of private firms in year $t-m$; AR_{t-m}^{PUB} = total applied research expenditure of public firms in year $t-m$; Y_{t-g}^{PV} = annual weighted yield index of private firms in year $t-g$; Y_{t-g}^{PUB} = annual weighted yield index of public firms at year $t-g$; $OY_{i,t-h}^{PV}$ = total yield index of private firms excluding firm i in year $t-h$; $OY_{j,t-h}^{PUB}$ = total yield index of public firms excluding firm j in year $t-h$; $u_{i,t}$, $u_{j,t}$ = random error terms, assumed to have multivariate normal with mean vector zero and covariance matrix Ω.

This model consists of a system of seven equations of seemingly unrelated regression: five for private firms and two for public firms. Some interesting practical features of the model are worth mentioning. First, each of the equations in the system contains its own lag of the dependent variable, so the system is dynamic. Secondly, given the limitation of the current data set, we have imposed cross-equation restrictions on both private and public firms. This enables us to estimate the parameters of the system adequately. Finally, we did not represent each equation with a general distributed lag model. We chose a simpler lag structure, looking for a single lag for each variable, assuming that it will take at least 4 years from basic research and 6 years from applied research for the first successful yield to be adopted. Indeed, we have tried to specify a more general system of autoregressive distributed lag model (SADL) and we did not find any significance for the recent lag structures.

The second and third specifications we consider are:

Model 2

$$R_{i,t}^{PV} = \alpha_{0i} + \alpha_1 AR_{i,t-k}^{PV} + \alpha_2 BR_{t-l} + \alpha_3 OAR_{i,t-r}^{PV}$$
$$+ \alpha_4 AR_{t-m}^{PUB} + \alpha_5 OY_{i,t-h}^{PV} + \alpha_6 Y_{i,t-g}^{PUB}$$
$$+ \alpha_7 HYB_{i,t}^{PV} + \alpha_8 PBR_t + \alpha_9 TUREV_{i,t}^{PV}$$
$$+ \lambda_i R_{i,t-1}^{PV} + u_{i,t}, \quad i = 1,\ldots,5 \qquad (6)$$

$$R_{j,t}^{PUB} = \theta_{0j} + \theta_1 AR_{j,t-k}^{PUB} + \theta_2 BR_{t-l} + \theta_3 OAR_{j,t-r}^{PUB}$$
$$+ \theta_4 AR_{t-m}^{PV} + \theta_5 OY_{j,t-h}^{PUB} + \theta_6 Y_{j,t-g}^{PV}$$
$$+ \theta_7 HYB_{i,t}^{PUB} + \theta_8 PBR_t + \theta_9 TUREV_{i,t}^{PUB}$$
$$+ \lambda_j R_{j,t-1}^{PUB} + u_{i,t}, \quad j = 1, 2 \qquad (7)$$

Model 3

$$SR_{i,t}^{PV} = \varphi_{0i} + \varphi_1 AR_{i,t-k}^{PV} + \varphi_2 BR_{t-l} + \varphi_3 OAR_{i,t-r}^{PV}$$
$$+ \varphi_4 AR_{t-m}^{PUB} + \varphi_5 PBR_t + \rho_i SR_{i,t-1}^{PV}$$
$$+ u_{i,t}, \quad i = 1,\ldots,5 \qquad (8)$$

$$SR_{j,t}^{PUB} = \mu_{0j} + \mu_1 AR_{j,t-k}^{PUB} + \mu_2 BR_{t-l}$$
$$+ \mu_3 OAR_{j,t-r}^{PUB} + \mu_4 AR_{t-m}^{PV} + \mu_5 PBR_t$$
$$+ \rho_j Y_{j,t-1}^{PUB} + u_{i,t}, \quad j = 1,2 \qquad (9)$$

where, we assume again, $|\gamma_i| < 1, |\gamma_j| < 1, |\rho_i| < 1$ and $|\rho_j| < 1$, for all i, j, and $R_{i,t}^{PV}$ = revenue of private firm i in year t; $R_{j,t}^{PUB}$ = revenue of public firm i in year t; $SR_{i,t}^{PV}$ = social revenue of private firm i in year t; $SR_{j,t}^{PUB}$ = social revenue of public firm i in year t; $HYB_{i,t}^{PV}$ = the proportion of the total area seeded to hybrid (HYB) varieties for private firm i at time t; $HYB_{j,t}^{PUB}$ = the proportion of the total area seeded to hybrid (HYB) varieties for public firm j at time t; PBR_t = plant breeders' rights dummy for private/public firm in year t; $TUREV_{i,t}^{PV}$ = TUA (technical use agreement) revenue for private firm i in year t; $TUREV_{i,t}^{PUB}$ = TUA (technical use agreement) revenue for public firm j in year t; and other variables defined previously as in model 1.

The specifications of models 2 and 3 are similar to those of model 1 in terms of lag structure specifications. For each model, the unknown parameters in the dynamic system, in principle, can be easily estimated by Zellner's iterative seemingly unrelated regression (ISUR) estimator. These estimates are consistent, asymptotically efficient and numerically equivalent to the maximum likelihood estimator.

The ISUR estimator uses equation-by-equation ordinary least squares (OLS) to construct an estimate of the disturbance covariance matrix Ω and then does the generalized least squares, given this initial estimate of Ω, on an appropriately stacked set of equations. The procedure is then iterated until the estimated parameters and the estimated Ω converge.

One estimation decision that arises in each model is how to choose the appropriate lag length. One simple way is to select the lag based on the minimum of the multivariate version of the Akaike information criterion (MAIC). Alternatively, given a special structure of the model, specifying different lags always results in the same number of parameters. Consequently, minimizing the MAIC is equivalent to minimizing the determinant of residual covariance matrix. We have used this second approach to determine the appropriate lag length in each model.

Regression results

The regression results for the three models reported in Tables 8.1, 8.2 and 8.3 appear to be robust. Most of the estimated coefficients are individually statistically significant at the 5% level. Almost all the explanatory variables have the

Table 8.1. Regression results of model 1.

Variable	Acronym		Coefficient	t-statistic	P
Private applied research expenditures in year t–6	$AR^{PV}_{i,t-k}$	β_1	2.116	8.171	0.000
Basic research expenditures in year t–9	BR_{t-l}	β_2	0.304	3.326	0.001
Total applied research expenditures of other private firms in year t–6	$OAR^{PV}_{i,t-r}$	β_3	0.320	3.813	0.000
Total applied research expenditures of public firms in year t–7	AR^{PUB}_{t-m}	β_4	0.158	1.831	0.069
Total yield index of private firms in year t–6	$OY^{PV}_{i,t-h}$	β_5	0.903	65.945	0.000
Yield index of public firms at year t–12	Y^{PUB}_{t-g}	β_6	−0.448	−5.600	0.000
Public applied research expenditures in year t–6	$AR^{PUB}_{j,t-k}$	δ_1	0.601	2.087	0.038
Basic research expenditures in year t–9	BR_{t-l}	δ_2	−0.200	−1.734	0.085
Total applied research expenditures of other public firms in year t–6	$OAR^{PUB}_{j,t-r}$	δ_3	0.351	3.320	0.001
Total applied research expenditures of private firms in year t–7	AR^{PV}_{t-m}	δ_4	−0.163	−2.018	0.045
Total yield index of other public firms in year t–6	$OY^{PUB}_{j,t-h}$	δ_5	0.036	2.297	0.023
Yield index of private firms at year t–12	Y^{PV}_{t-g}	δ_6	0.000	−0.317	0.752
Yield index of private firm 1 in year t–1	$Y^{PV}_{1,t-1}$	γ_1	0.067	4.221	0.000
Intercept private firm 1	Constant	β_{01}	37.665	5.128	0.000
Yield index of private firm 2 in year t–1	$Y^{PV}_{2,t-1}$	γ_2	0.335	3.761	0.000
Intercept private firm 2	Constant	β_{02}	54.000	4.927	0.000
Yield index of private firm 3 in year t–1	$Y^{PV}_{3,t-1}$	γ_3	0.479	6.419	0.000
Intercept private firm 3	Constant	β_{03}	48.706	4.758	0.000
Yield index of private firm 4 in year t–1	$Y^{PV}_{4,t-1}$	γ_4	0.500	7.431	0.000
Intercept private firm 4	Constant	β_{04}	42.836	4.491	0.000
Yield index of private firm 5 in year t–1	$Y^{PV}_{5,t-1}$	γ_5	0.500	6.697	0.000
Intercept private firm 5	Constant	β_{05}	47.323	4.712	0.000
Yield index of public firm 1 in year t–1	$Y^{PUB}_{1,t-1}$	γ_6	0.521	5.251	0.000
Intercept public firm 1	Constant	δ_{06}	50.287	4.858	0.000
Yield index of public firm 2 in year t–1	$Y^{PUB}_{2,t-1}$	γ_7	0.940	14.495	0.000
Intercept private firm 2	Constant	δ_{07}	−0.615	−0.131	0.896

Dependent variables: annual weighted yield index of private firms i in year t: $Y^{PV}_{i,t}$, public firms: $Y^{PUB}_{i,t}$. Determinant residual covariance: $1.71E + 12$. $\bar{R}^2$: 0.590–0.997.

Table 8.2. Regression results of model 2.

Variable	Acronym		Coefficient	t-statistic	P
Private applied research expenditures in year t–9	$AR^{PV}_{i,t-k}$	α_1	0.480	1.854	0.065
Basic research expenditures in year t–7	BR_{t-l}	α_2	0.346	2.777	0.006
Total applied research expenditures of other private firms in year t–9	$OAR^{PV}_{i,t-r}$	α_3	−0.341	−1.852	0.066
Total applied research expenditures of public firms in year t–8	AR^{PUB}_{t-m}	α_4	0.311	2.725	0.007
Total yield index of other private firms in year t–9	$OY^{PV}_{i,t-h}$	α_5	−0.309	−6.477	0.000
Yield index of public firms at year t–12	Y^{PUB}_{t-g}	α_6	−0.305	−2.663	0.009
The proportion of the total area seeded to hybrid (HYB) varieties for private firm at time t	$HYB^{PV}_{i,t}$	α_7	3.466	2.678	0.008
Plant breeders' right dummy for private/public firm in year t	PBR_t	α_8	5.592	2.992	0.003
TUA (technical use agreement) revenue for private firm in year t	$TUREV^{PV}_{j,t}$	α_9	0.943	11.966	0.000
Public applied research expenditures in year t–9	$AR^{PUB}_{j,t-k}$	θ_2	0.962	3.231	0.002
Basic research expenditures in year t–7	BR_{t-l}	θ_1	−0.187	−0.639	0.524
Total applied research expenditures of other public firms in year t–9	$OAR^{PUB}_{j,t-r}$	θ_3	−2.412	−4.159	0.000
Total applied research expenditures of private firms in year t–8	AR^{PV}_{t-m}	θ_4	0.278	1.050	0.295
Total yield index of other public firms in year t–9	$OY^{PUB}_{j,t-h}$	θ_5	0.247	1.816	0.071
Yield index of private firms at year t–12	Y^{PV}_{t-g}	θ_6	0.00022	−2.740	0.007
The proportion of the total area seeded to hybrid (HYB) varieties for public firm at time t	$HYB^{PUB}_{j,t}$	θ_7	−3.996	−0.842	0.401
Plant breeders' right dummy for private/public firm in year t	PBR_t	θ_8	−8.140	−1.628	0.105
TUA (technical use agreement) revenue for public firm in year t	$TUREV^{PV}_{j,t}$	θ_9	7.738	1.393	0.165
Revenue of private firm 1 in year t–1	$R^{PV}_{1,t-1}$	λ_1	1.199	7.258	0.000
Intercept private firm 1	Constant	α_{01}	24.777	2.377	0.019
Revenue of private firm 2 in year t–1	$R^{PV}_{2,t-1}$	λ_2	0.412	3.763	0.000
Intercept private firm 2	Constant	α_{02}	25.347	2.434	0.016
Revenue of private firm 3 in year t–1	$R^{PV}_{3,t-1}$	λ_3	0.882	14.516	0.000
Intercept private firm 3	Constant	α_{03}	22.105	2.115	0.036

Continued

Table 8.2. *Continued.* Regression results of model 2.

Variable	Acronym		Coefficient	t-statistic	P
Revenue of private firm 4 in year t–1	$R^{PV}_{4,t-1}$	λ_4	0.497	6.137	0.000
Intercept private firm 4	Constant	α_{04}	25.885	2.492	0.014
Revenue of private firm 5 in year t–1	$R^{PV}_{5,t-1}$	λ_5	0.636	10.483	0.000
Intercept private firm 5	Constant	α_{05}	25.844	2.484	0.014
Revenue of public firm 1 in year t–1	$R^{PUB}_{1,t-1}$	λ_6	0.437	3.428	0.001
Intercept public firm 1	Constant	θ_{06}	41.328	4.419	0.000
Revenue of public firm 2 in year t–1	$R^{PUB}_{2,t-1}$	λ_7	0.383	2.714	0.007
Intercept private firm 2	Constant	θ_{07}	−16.897	−1.160	0.248

Dependent variable: revenue of private firm i in year t: $R^{PV}_{i,t}$, public firm j in year t: $R^{PUB}_{j,t}$.
Determinant residual covariance: 6.83E+08, $\overline{R}^2$: 0.467–0.963.

expected signs. The regressions have $\overline{R}^2$ between 0.590 and 0.997 (first regression), 0.467 and 0.963 (second regression), and 0.342 and 0.890 (third regression).

Model 1

The firms' own-lagged applied research expenditure has a positive effect on yield. The coefficient of 2.12 for private firms (0.601 for public firms) implies that a $1 million[3] expenditure increases the yield index by 2.12 (0.601). The much larger coefficient for the private firms suggests a higher direct productivity for private applied research. For all firms, public and private, the previous years' yields have positive signs, with coefficients less than 1, and thus are consistent with dynamic stability.

The empirical results reveal that lagged basic research expenditure positively affects the annual weighted yield index of private firms, while negatively affecting the weighted yield index of public firms. Public basic research expenditure with a lag of nine periods has a coefficient of 0.304 in the first model, implying that, *ceteris paribus*, a $1 million increase in the annual public basic research in 1 year increases the private yield index after 9 years by 0.304 index points. This positive spillover is consistent with the notion that basic research increases the productivity of private applied research. In contrast to this result, a $1 million increase in the annual public basic research expenditure in 1 year reduces the public yield index

after 9 years by 0.2 index points. This interesting result suggests that an increase in basic research, which is located within public institutions, uses common resources within the research institution, thereby reducing the resources available for applied public research.

Other firms' lagged research expenditures have a spillover effect on each firm's yield index. The synergistic effect was strongest within groups, i.e. between public firms (0.35) and between private firms (0.32). A somewhat smaller synergistic effect was evident between groups in the spillover of public expenditure on applied yields (0.158). These positive effects are consistent with human capital and knowledge spillovers. A negative spillover effect of 0.163 occurred between private firm expenditures and public firm yields. This latter between-group effect may have been generated from private firms bidding highly qualified personnel away from the public sector. During the growth phase of the industry, migration tended to occur from the public sector to the private sector.

A positive spillover was evident for yields within groups, while the spillover was negative between groups. A one point increase in other private (public) firms' yield index resulted in a 0.9 point (0.036) increase in the firm's own yield index. In contrast, the public yield index had a negative 0.448 point impact on private yield, while the reverse between-group impact was also negative but insignificant.

Table 8.3. Regression results of model 3.

Variable	Acronym		Coefficient	t-statistic	P
Private applied research expenditures in year t–11	$AR^{PV}_{i,t-k}$	φ_1	1.846	1.730	0.085
Basic research expenditures in year t–7 φ_1	BR_{t-1}	φ_2	0.806	2.273	0.024
Total applied research expenditures of other private firms in year t–8	$OAR^{PV}_{i,t-r}$	φ_3	−1.962	−3.774	0.000
Total applied research expenditures of public firms in year t–13	AR^{PUB}_{t-m}	φ_4	1.067	2.284	0.024
Plant breeders' right dummy in year t	PBR_t	φ_5	29.947	5.083	0.000
Public applied research expenditures in year t–11	$AR^{PUB}_{j,t-k}$	μ_1	5.236	2.981	0.003
Basic research expenditures in year t–7	BR_{t-1}	μ_2	−5.727	−3.173	0.002
Total applied research expenditures of other public firms in year t–8	$OAR^{PUB}_{j,t-r}$	μ_3	−6.243	−3.117	0.002
Total applied research expenditures of private firms in year t–13	AR^{PV}_{t-m}	μ_4	2.915	1.215	0.226
Plant breeders' right dummy in year t	PBR_t	μ_5	−42.473	−1.851	0.066
Social revenue of private firm 1 in year t–1	$SR^{PV}_{1,t-1}$	ρ_1	0.764	7.001	0.000
Intercept private firm 1	Constant	φ_{01}	−3.569	-0.634	0.527
Social revenue of private firm 2 in year t–1	$SR^{PV}_{2,t-1}$	ρ_2	0.221	2.043	0.043
Intercept private firm 2	Constant	φ_{02}	−4.157	-1.163	0.246
Social revenue of private firm 3 in year t–1	$SR^{PV}_{3,t-1}$	ρ_3	0.726	9.972	0.000
Intercept private firm 3	Constant	φ_{03}	0.088	0.013	0.990
Social revenue of private firm 4 in year t–1	$SR^{PV}_{4,t-1}$	ρ_4	0.697	9.390	0.000
Intercept private firm 4	Constant	φ_{04}	−14.000	-1.286	0.200
Social revenue of private firm 5 in year t–1	$SR^{PV}_{5,t-1}$	ρ_5	0.697	8.560	0.000
Intercept private firm 5	Constant	φ_{05}	−1.597	-0.292	0.771
Social revenue of public firm 1 in year t–1	$SR^{PUB}_{1,t-1}$	ρ_6	0.425	3.377	0.001
Intercept public firm 1	Constant	φ_{06}	210.067	5.205	0.000
Social revenue of private firm 2 in year t–1	$SR^{PUB}_{2,t-1}$	ρ_7	0.587	4.642	0.000
Intercept private firm 2	Constant	φ_{07}	62.267	3.526	0.001

Dependant variables: social revenue of private firm in year t: $SR^{PV}_{i,t}$, public firm: $SR^{PUB}_{j,t}$.
Determinant residual covariance: 3.63E+18, $\bar{R}^2$: 0.342–0.890.
Source: Authors' regression estimates.

The results of model 1 show that a firm's current yield index can be modelled as a function of previous research expenditure. The model revealed strong evidence of positive spillovers within the public and the private sectors. Publicly funded basic research and applied research created a positive spillover for private yields. Other public/private spillovers were negative in sign.

Model 2

Model 2, which examines the determinants of firm revenue, revealed that $1 of own firm lagged applied research increased private (public) revenue by $0.480 ($0.962). This model also showed important spillover effects. In this case, the spillovers include pecuniary effects in the output market and therefore illuminate crowding effects. An additional dollar in lagged basic research expenditure changed private (public) revenue by $0.346 (−$0.187), indicating that public basic research provides monetary benefits to private industry, while drawing resources away for public firm applied research.

The inter-firm spillover effects of lagged applied research were negative within groups. A $1 increase in other private (public) firm applied research expenditure reduced firm revenue by $0.341 ($2.412). Given that there were positive spillovers in production, these negative impacts show a strong degree of competition within groups, which is not surprising since the firms are competing for the same customers.

In contrast to the within-group competition, a $1 increase in public (private) expenditure increased private revenue by $0.311 ($0.278), indicating positive spillovers between groups. This indicates that non-pecuniary spillovers dominate the pecuniary spillovers such that public applied research activity has crowded in private research, rather than crowding it out.

The spillover of other firms' yields tends to have a negative impact on firm revenue. This negative relationship exists among private firms, from private to public firms, and from public to private firms. The exception is the public-to-public interaction, where there is synergistic impact, perhaps due to a different ethos among public breeders.

The variables for proportion of the total area seeded to hybrids and for PBRs had a positive impact on private revenues, while having a negative impact on public revenue. A complete shift to hybrids would increase (reduce) private (public) revenue by $3.466 million ($3.996 million) per year. PBRs increased (reduced) private (public) revenue by $5.592 million ($8.14 million). The TUA fees had a positive effect on total revenue; 0.94 in the case of private firms, suggesting a slight reduction in the non-TUA revenue, while for the public firms $1 in TUA revenue tended to increase total revenue by $7.738, indicating a dramatic increase in pricing.

In summary, model 2, which examines firm revenue, shows evidence of the pecuniary impacts of competition between firms, particularly within groups. Applied within-group expenditure reduces other firm revenue, while between-group spillovers are positive. A higher lagged yield for competing firms has a negative impact on revenue, with the exception of public-to-public impacts, for which it is positive. Property rights and hybrid technologies have a positive effect on private sales revenue and a negative impact on public revenue.

Model 3

The estimates of model 3 show how the social revenue associated with the varieties of each firm are affected by research expenditures and PBRs. The results are similar in sign to the private revenue estimated in model 2.

The applied research investment in each firm increases the social revenue associated with its varieties. In the case of private (public) firms, a $1 increase in applied research resulted in an increase in social revenue of $1.846 ($5.236). These figures are much larger than the increase in private revenue reported in model 2, indicating a gap between private and social revenue and a significant positive spillover to downstream research users, particularly in the case of public applied research.

A $1 increase in lagged basic research increased (reduced) private (public) social revenue by $0.806 ($5.727). This indicates that the output of private firms is positively affected by basic research; the public variety output once again is a decreasing function of public basic research expenditure.

The other firms' research expenditure has a negative impact within group and a positive impact between groups.

A $1 increase in a private (public) firms competitor's applied research reduced the firm's associated social revenue by $1.962 ($6.243).

An increase in private (public) applied research increased the social revenue associated with public (private) varieties by $1.06 ($2.195). Comparing the significantly smaller own research impacts with the larger negative spillovers would suggest that an increase in applied research could have a negative impact on social revenue.

PBRs have a strong positive effect on the impact of private research and a strong negative impact on the products of public applied research. The estimates suggest that PBRs increased the social revenue associated with private varieties by $29.95 million while reducing revenue associated with public applied research by $42.47 million. This is a very substantial shift and probably reflects other changes in research policy that coincided with PBRs, including the introduction of the practice of transferring public varieties to public firms for commercialization.

In summary, model 3 shows that the social revenue associated with the output of a firm can be estimated as a function lagged research expenditure and PBRs. The results are consistent with model 2 and show that competition within a group is much stronger than that between groups. The fact that the estimated coefficients for social revenue from the own-applied research are greater than the private revenue coefficients suggests a significant spillover of benefits to downstream research users. The fact that the across-firm negative spillovers from applied research are greater than the positive own firm effects suggests there could be overexpenditure in the industry. The introduction of PBRs coincided with a major change in the social revenue associated with private and public varieties.

Conclusions

This study examined many research spillovers in a modern crop research industry as delineated by: their public or private source; their public or private incidence; whether they were generated through basic research, applied research activity or germplasm; and whether they were inter-firm pecuniary, inter-firm non-pecuniary or downstream in nature. The empirical framework, which estimated a production function, a private revenue function and a social revenue, provided a useful conceptual separation of research spillovers, and provided a broad scope of empirical results with

many implications for private incentives and research policy.

The three empirical models fit the data well and provided theoretically plausible estimates. Lagged applied research investment by each firm increased research output, research revenue and social research revenue. Enhanced IPRs increased private research revenue, and the social value of their innovations. Perhaps the most striking general result was the ubiquitous presence of research spillovers in each model.

The empirical results of model 1 provide the strongest evidence of non-pecuniary spillovers. The results show that public basic research, public applied research, other private firm applied research and other private firm varieties created a positive spillover for private firms. These spillovers indicate that public research has made private research firms more productive, and private firms may benefit from the knowledge generated from their rivals.

The empirical results of model 2 provide estimates of the total (pecuniary plus non-pecuniary) research spillovers. The results show that, while private firms have a net competitive or crowding out effect, public firm basic and applied research enhances private revenue, creating a crowding in effect. This model also shows that PBRs, proprietary technologies and TUAs enhance private firms' ability to generate revenue.

Model 3 empirically estimates the social value of sales from each firm. The much larger coefficients than those estimated in model 2 provide evidence of considerable downstream research spillovers in the oilseed rape industry. Spillovers from public basic and applied research have enhanced the social value of firm output. The large impact of the PBRs represents a significant structural change that significantly increased the social value of private firms' output while reducing the social value of sales from public institutions.

The results of the empirical analysis have several implications for research policy. The most apparent is that public and private research firms are integrally linked through numerous types of research spillovers. Publicly funded basic and applied research both had positive effects on private research productivity, profitability and social value output. The negative impact of basic research on public firm output and revenue suggests that these basic research activities are under-reported and tend to use resources earmarked for applied research. Given the importance of basic research to private industry output, this diversion of resources could be optimal. The ability of public institutions to conduct applied research while crowding in private applied research suggests that public policies such as the Matching Investment Initiative have been successful in mitigating the normal crowding effects. The positive impact that IPRs had on private revenue suggests that these changes have been effective in providing incentives for private research.

The prevalence of non-pecuniary inter-firm research spillovers suggests a strong research clustering effect – an effect that is particularly evident in Saskatoon where there is a significant concentration of public and private firms involved in oilseed rape research. The existence of a clustering effect suggests the need for a mechanism for the coordination of private and public location choices to maximize the spillover opportunities. The significant public-to-private spillovers emphasize the importance of the public institutions in these clusters.

In this study, we found empirical evidence of a variety of research spillovers in the oilseed rape research industry. The importance of research to economic growth suggests a need to understand these complex non-market relationships fully, and to manage research policy with these spillovers in mind.

Notes

[1] We were unable to find any data reporting the amount of basic research undertaken by private firms and therefore cannot estimate this effect. Industry experts indicated that this activity was very limited in private firms.

[2] The input use data were collected in terms of the number of scientist years per firm. This was converted to real dollar terms using the estimated 2003 $/scientist ratio to facilitate easier comparison with the equations estimated in models 2 and 3.

[3] All values are given in Canadian dollars.

References

Acs, Z.J., Audretsch, D.B. and Feldman, M.P. (1994) R&D spillovers and recipient firm size. *Review of Economics* 76, 336–340.

Adams, J.D. (1990) Fundamental stocks of knowledge and productivity growth. *Journal of Political Science* 98, 673–702.

Adams, J.D. (2000) *Endogenous R&D Spillovers and Industrial Research Productivity.* Working Paper 7484. National Bureau of Economic Research, Cambridge, Massachusetts.

Aghion, P. and Howitt, P. (1992) A model of growth through creative destruction. *Econometrica* 60, 323–351.

Alberta Agriculture and Food (2002, 2003) *2003 Agronomic Performance Data for all Crops and Regions.* http://www.agric.gov.ab.ca/crops/performance/all.html

Alberta Crop Insurance Corporation (2002) Personal communication with Murray Hartman, Extension Agronomist, Alberta Agriculture and Food. December.

Alberta Seed Industry (2001) *New in 2001: Canola, Flax, and Mustard Varieties – 2001.* http://www.seed.ab.ca/var/sum/gmcanola01.html

Alston, J.M. and Pardey, P.G. (1996) *Making Science Pay: the Economics of Agricultural R&D Policy.* AEI Press, Washington, DC.

Bergstrom, T., Blume, L. and Varian, H. (1986) On the private provision of public goods. *Journal of Political Economy* 29, 25–49.

Canada Department of Justice (2000) http://laws.justice.gc.ca./en/P-14.8/index.html

Canadian Food Inspection Agency (1998) *Canadian Varieties. January 1, 1923 to June 24, 1998.* Special Tabulation from the Plant Health and Production Division. http://inspection.gc.ca

Canadian Food Inspection Agency (2002, 2003) *Status of PBR Applications and Grant of Rights: Complete List of Varieties.* http://inspection.gc.ca/english/plaveg/pbrpov/ croproe.pdf

Canadian Food Inspection Agency (2002) http://inspection.gc.ca/english/plaveg/variet/rapecole.shtml

Canola Council of Canada (2002,2003) *Argentine Varieties and Polish Varieties.* http://www.canola-council.org/production/2_varint.html

Canola Council of Canada (2002) *Economic Analysis: CPC Annual Reports* (Various Years). http://www.canola-council.org/(1998–2001). From 1991 to 1997, obtained through personal communication with Dave Wilkins, Director of Communications, Canola Council of Canada.

Canola Research Survey (1999) University of Saskatchewan, Department of Agricultural Economics, Saskatoon, Saskatchewan.

Cohen, W.M. and Levinthal, D.A. (1989) Innovation and learning: the two faces of R&D. *Economic Journal* 99, 569–597.

Dasgupta, P. and Stiglitz, J. (1980) Industrial structure and the nature of innovative activity. *Economic Journal* 90, 266–293.

David, P.A. and Hall, B.H. (2000) *Heart of Darkness: Modeling Public–Private Funding*

Interactions Inside the R&D Black Box. Report prepared for a special issue of *Research Policy.* National Bureau of Economic Research, Cambridge, Massachusetts.

David, P.A., Hall, B.H. and Toole, A.A. (1999) *Is Public R&D a Complement or Substitute for Private R&D? A Review of the Econometric Evidence.* Report prepared for a special issue of *Research Policy.* National Bureau of Economic Research, Cambridge, Massachusetts.

Diamond, A.M., Jr (1999) Does federal funding 'crowd in' private funding of science? *Contemporary Economic Policy* 17, 423–431.

Evenson, R.E. (1989) Spillover benefits of agricultural research: evidence from US experience. *American Journal of Agricultural Economics* 71, 447–452.

Evenson, R.E. (1999) Intellectual property rights, access to plant germplasm, and crop production scenarios in 2020. *Crop Science* 39, 1630–1635.

Evenson, R.E. and Kislev, Y. (1976) A stochastic model of applied research. *Journal of Political Economy* 84, 265–281.

Falcon W.P. and Fowler, C. (2002) Carving up the commons – emergence of a new international regime for germplasm development and transfer. *Food Policy* 27:197–222. www.elsevier.com/locte/foodpol

Fernandez-Cornejo, J. (2004) The seed industry in U.S. agriculture: an exploration of data and information on crop seed markets, regulation, industry structure, and research and development. Resource Economics Division, Economic Research Service, US Department of Agriculture. *Agriculture Information Bulletin* 786, February, 2004.

Foster, A.D. and Rosenzweig, M.R. (1995) Learning by doing and learning from others: human capital and technical change. *Journal of Political Economy* 103, 1176–1209.

Fulton, M. (1997) The economics of intellectual property rights: discussion. *American Journal of Agricultural Economics* 79, 1592–1594.

Fulton, M. and Giannakas, K. (2001) Agricultural biotechnology and industry structure. *AgBioForum* 4, 137–151.

Glaeser, E.L., Kallal, H.D., Sheinkman, J. and Scheifer, A. (1992) Growth in cities. *Journal of Political Economy* 100, 1126–1152.

Graff, G.D., Cullen, S.E., Bradford, K.J., Zilberman, D. and Bennet, A.B. (2003) The public–private structure of intellectual property ownership in agricultural biotechnology. *Nature Biotechnology* 21, 989–995.

Gray, R.S., Malla, S. and Phillips, P. (2002) Gains to yield increasing research in the evolving Canadian canola research industry. In: Santaniello, V., Evenson, R.E. and Zilberman, D. (eds) *Market Development for Genetically Modified Foods.* CAB International, Wallingford, UK, pp. 113–126.

Griliches, Z. (1979) Hybrid corn revisited: a reply. *Econometrica* 48, 1463–1465.

Griliches, Z. (1980) Issues in assessing the contribution of research and development to productivity growth. *Bell Journal of Economics* 10, 92–116.

Griliches, Z. (1992) The search for R&D spillovers. *Scandinavian Journal of Economics* 94 (Supplement), 47.

Heisey, P. and Morris, M.L. (2002) *Practical Challenges to Estimating the Benefits of Agricultural R&D: the Case of Plant Breeding Research.* Selected paper for presentation at the American Agricultural Economic Association Meetings, Long Beach, California, July 2002.

Huffman, W.E. and Evenson, R.E. (1993) *Science for Agriculture: a Long-Term Perspective.* Iowa State University Press, Ames, Iowa.

Institute for Scientific Investigation (1997) Citations Database, special tabulation of academic publications based on keyword search for 'canola'. November, 1997.

Inventory of Canadian Agri-Food Research (1998) Special tabulation upon request. http://res2.agr.ca/icar/english/tableofcontents.htm

Inventory of Canadian Agri-Food Research (2000) Special tabulation upon request. http://res2.agr.ca/icar/english/tableofcontents.htm

Jaffe, A.B. (1986) Technological opportunity and spillovers of R&D: evidence from firm's patents, profits, and market value. *American Economic Review* 76, 984–1001.

Johnson, D.K.N. and Evenson, R.E. (1999) R&D spillovers to agriculture: measurement and application. *Contemporary Economic Policy* 17, 432–456.

Just, R.E. and Hueth, D.L. (1993) Multimarket exploitation: the case of biotechnology and chemicals. *American Journal of Agricultural Economics* 75, 936–945.

Khanna, J. and Sandler, T. (1996) *UK Donations Versus Government Grants: Endogeneity and Crowding-in.* Presented at American Economic Association meeting, San Francisco, California, January 1996.

Khanna, J., Posnett, J. and Sandler, T. (1995) Charity donations in the UK: new evidence based on panel data. *Journal of Political Economy* 56, 257–272.

Kortum, S. (1997) Research, patenting, and technological change. *Econometrica* 65, 1389–1419.

Lee, T.K. (1982) A nonsequential R&D search model. *Management Science* 28, 900–909.

Lee, T.K. (1985) On the joint decisions of R&D and technology adoption. *Management Science* 31, 959–969.

Lesser, W. (1997) Assessing the implications of intellectual property rights on plant and animal agriculture. *American Journal of Agricultural Economics* 79, 1584–1591.

Lesser, W. (1998) Intellectual property rights and concentration in agricultural biotechnology. *AgBioForum* 1, 56–61.

Levin, R.C. and Reiss, P.C. (1984) Test of Schumpeterian model of R&D and market structure. In: Griliches, Z. (ed.) *R&D, Patents, and Productivity*. University of Chicago Press, Chicago, Illinois, pp. 175–204.

Levin, R.C. and Reiss, P.C. (1988) Cost-reducing and demand-creating R&D with spillovers. *RAND Journal of Economics* 19, 538–556.

Lucas, R. (1988) On the mechanics of economic development. *Journal of Monetary Economics* 22, 3–42.

Malla, S., Gray, R. and Phillips, P. (2004) Gains to research in the presence of IPRs and research subsidies. *Review of Agricultural Economics* 26, 63–81.

Manitoba Agriculture and Food (2003) *2003 Variety Guide for Oilseeds and Special Crops*. http://www.gov.mb.ca/agriculture/crops/oilseeds/pdf/bga01s08.pdf

Manitoba Crop Insurance Corporation (2002) Manitoba's Management Plus Program. *Regional Variety Yield Analysis (1991–2001)*. http://www.mmpp.com/Information_Page.htm

Moschini, G. and Lapan, H. (1997) Intellectual property rights and the welfare effects of agricultural innovation. *American Journal of Agricultural Economics* 79, 1229–1242.

Nagy, J.G. and Furtan, W.H. (1977) *The Socio-economic Costs and Returns from Rapeseed Breeding in Canada*. University of Saskatchewan, Department of Agricultural Economics, Saskatoon, Saskatchewan.

Nagy, J.G. and Furtan, W.H. (1978) Economic costs and returns from crop development research: the case of rapeseed breeding in Canada. *Canadian Journal of Agricultural Economics* 26, 1–14.

Pardey, P.G., Alston, J.M., Christian, J.E. and Fan, S. (1996) *Hidden Harvest: U.S. Benefits from International Research Aid*. Food Policy Report. International Food Policy Research Institute (IFPRI), Washington, DC.

Paterson, N.M. and Sons Ltd Grain Buyer (2003) *Polish and Argentine Canola Varieties*. http://www.patersongrain.com/cropinputs.html

Phillips, P. (1997) Manual search of ISI Citations Index, July.

Prairie Pools Inc. Various Issues. *Prairie Grain Variety Survey 1977–1992*. Regina, Saskatchewan.

Roberts, R.D. (1984) A positive model of private charity and public transfers. *Journal of Political Economy* 92, 136–148.

Robson, M. (1993) Federal funding and the level of private expenditure on basic research. *Southern Economic Journal* 60, 63–71.

Romer, P. (1986) Increasing returns and long-run growth. *Journal of Political Economy* 94, 1002–1037.

Romer, P. (1990) Endogenous technological change. *Journal of Political Economy* 98, S71–S102.

Saskatchewan Agriculture and Food (2002) *Agricultural Statistics*. Saskatchewan Agriculture and Food, Regina, Saskatchewan.

Saskatchewan Agriculture and Food. Various Issues. *Varieties of Grains Crops in Saskatchewan*. Saskatchewan Agriculture and Food, Regina, Saskatchewan.

Schimmelpfennig, D., Pray, C.E. and Brennan, M.F. (2004) The impact of seed industry concentration on innovation: a study of US biotech market leaders. *Agricultural Economics* 30, 157–167.

SeCan (2002) Prices, royalties and volumes. Personal communication with L.R. White, General Manager, SeCan, Ottawa, Ontario, November.

Schultz, T.W. (1975) The value of the ability to deal with disequilibria. *Journal of Economic Literature* 13, 827–846.

Spence, A.M. (1984) Cost reduction, competition, and industry performance. *Econometrica* 52, 1001–1021.

Statistics Canada (2002) *Direct Cansim Time Series: Prairie Provinces; Seeded Area; Canola (Rapeseed)*. http://datacenter2.chass.utoronto.ca/cgi-bin/cansim2/getSeriesData.pl?s=V169455&b=&e=&f=plain

Statistics Canada (2003) *Direct Cansim Time Series: CPI and All Goods for Canada. Direct Cansim Time Series: CPI and All Goods for Canada*. http://datacenter2.chass.utoronto.ca/cgi-bin/cansim2/getSeriesData.pl?s=V735319&b=&e=&f=plain

Thorton, R.A. and Thompson, P. (2001) Learning from experience and learning from others: an exploration of learning and spillovers in wartime shipbuilding. *American Economic Review* 91, 1350–1368.

White, F.C., He, S. and Fletcher, S. (2003) Research spillovers and returns to wheat research investment. Selected paper for presentation at the *Southern Agricultural Economics Association Annual Meeting*, Mobile, Alabama, February 2003.

Yee, J. and Huffman, W. (2001) *Rates of Return to Public Agricultural Research in the Presence of Research Spillovers*. Presented at the American Agricultural Economic Association Meetings, Chicago, Illinois, August 2001.

Appendix 8.1: Data Description

This Appendix describes the source and the calculations used to construct each of the variables required for the econometric analysis.

Research expenditure on oilseed rape

Developing the time series for public and private research expenditures from 1960 to 1999 required the combination of several sources of data and several calculations because no single source spanned the time period and some sources were more accurate than others for some types of expenditures. For all of the calculations, the total research expenditure per year was calculated by multiplying the total (professional and technical) person-years employed in research each year by the 1999 total research costs per person (Canola Research Survey, 1999). To avoid the problem of double counting, the research expenditure is calculated at the final recipient level. Data on oilseed rape research person-years were obtained from five sources: Canola Research Survey (1999); Nagy and Furtan (1977, 1978); Institute for Scientific Investigation (1997); Phillips (1997); and Inventory of Canadian Agri-Food Research (1998, 2000).

Canadian universities' total person-years were based on an Institute for Scientific Investigation (ISI) special tabulation of academic publications (from 1981 to 1996) and Inventory of Canadian Agri-Food Research (ICAR) data (from 1977 to 1980) (for the 1978 value, the average of 1977 and 1979 was used). Prior to 1976, when ICAR data were not available, Nagy and Furtan (1977, 1978) was used, with some adjustments. Comparing the Nagy and Furtan estimates of total professional person-years at Canadian universities with the ICAR estimates of total professional years, the former were underestimated by 62%. Hence,

the Nagy and Furtan estimates were adjusted by multiplying by 2.64. The data were updated by applying the average value of the last three available years to the 1997, 1998 and 1999 values.

Non-Canadian universities' total person-years were based on an ISI special tabulation of academic publications (from 1981 to 1997), and on Phillips (1997) (from 1960 to 1980). Comparing the ISI special tabulation of non-Canadian academic total person-years with the Phillips' estimate (for the overlapping years 1981 and 1982), the former was underestimated by 23%. Hence, Phillips' estimates were adjusted by multiplying by 1.3. The non-Canadian universities' total person-years figure was updated following the same methodology as the Canadian universities' total person-years.

AAFC total person-years were based on ICAR data (1998 and 2000) from 1977 to 1999 (for the 1978 value, the average of 1977 and 1979 was used), and Nagy and Furtan (1977, 1978) (from 1960 to 1976). Comparing the Nagy and Furtan estimate of professional person-years with the ICAR estimate of total professional years, the former represented only 30% of the total person-years. Hence, the Nagy and Furtan (1977, 1978) estimates were adjusted accordingly.

The estimation of public expenditures on basic and applied research was based on the ICAR database and the public research expenditure (as described above). Upon request, ICAR personally provided us with project descriptions and subcategories of research in the 556 projects undertaken over the years on oilseed rape research. With the help of experts in crop science, the research in each project was divided into basic and applied research, and then aggregated to calculate the percentage of basic and applied research in each year. This percentage for the ICAR-listed projects was applied to all reported public research expenditures on rape, which resulted in a time series of public expenditures on basic and on applied research.

The estimation of the private companies' professional years was based on the Canola Research Survey (1999), a detailed firm-level study undertaken by Peter Phillips and others at the University of Saskatchewan.

Yield index

The annual yield index by firm was created from an average of the yield index for the firm's varieties grown each year, weighted by the seeded acreage.[i] The relative yields of different rape varieties were obtained from various issues of Saskatchewan Agriculture and Food, *Varieties of Grain Crops in Saskatchewan*, which are based on annual side-by-side variety yield trails at several locations in the province.[ii] The data on the percentage of acreage sown to each variety were obtained from four sources: Nagy and Furtan (1978); various issues of Prairie Pools Inc. (1977–1992); and the authors' estimates based on Manitoba Crop Insurance Corporation (2002) and Alberta Crop Insurance Corporation (2002).[iii]

Variety classification

Information on the types of rapeseed varieties (Polish/Argentine), the breeders, the year of introduction, the variety's reproduction system (open-pollinated, synthetic, hybrids), the variety's production system (conventional versus herbicide-tolerant, e.g. Roundup Ready®, Clearfield®, Liberty®), which allowed us to classify each oilseed rape variety, was collected from the following sources: Saskatchewan Agriculture and Food, *Varieties of Grain Crops in Saskatchewan* (various issues); Canadian Food Inspection Agency (1998, 2002, 2003); Alberta Agriculture and Food (2002, 2003); Manitoba Agriculture and Food (2003); Paterson and Sons Limited Grain Buyer (2003); Alberta Seed Industry (2003); and Canola Council of Canada (2002, 2003).

Argentine and biotechnology variables (herbicide-tolerant (HT); hybrid (HYB); plant breeders' rights (PBRs))

To capture the effect of cultivating herbicide-tolerant (HT) and synthetic/hybrid rape varieties (HYB), a variable was created that shows the proportion of the total oilseed rape area seeded to varieties that are either HT or hybrids. The HYB variables take a value between 0 and 1 (for details on the sources of varieties' classification see above).

The effect of plant breeders' rights (PBRs) was incorporated by creating a PBR dummy variable. This variable takes the value of 0 before the PBR act came into force on 1 August, 1990 (Canada Department of Justice, 2000), and 1 thereafter.

Price and revenue variables (private revenue; technical use agreement (TUA) revenue; social revenue)

The farm gate price of rapeseed in Canada was based on Saskatchewan Agriculture and Food (2002). The farm gate price as well as all the revenue variables were expressed as 2001 Canadian dollars/t as deflated by the consumer price index. The data for the consumer price index (CPI) were obtained from Statistics Canada (2003).

The annual revenue by firm is the product of the price charged for each variety and the seeded acreage for each variety (for details on the sources of the area data see, 'Yield index'). The price data of varieties were obtained from various issues of Canola Council of Canada (2002);[iv] SeCan (2002); Saskatchewan Agriculture and Food (2002); and the authors' estimates based on the above sources.[v]

The technical use agreement (TUA) fees or their annual rents equivalent were obtained from various issues of Canola Council of Canada (2002) and authors' estimates.[vi] The TUA revenue was calculated by multiplying the TUA fees/equivalent and the area seeded per variety per year.

Finally, the estimation of the annual social revenue produced by each firm each year was based on the notion that social value can be broken down to yield-induced increase and the herbicide cost savings reflected in the TUA fees. The estimated value of the yield increase from firm i in year t begins with the calculation of the commercial value of the product grown to their varieties, which is the product of the area A_{it}, the price of rape, P_t, and the average commercial yield in year t, Yt. The social value of the yield increase attributed to firm i is this commercial value multiplied by the proportional yield increase over the 1960 yield, or $(Y_{it} - Y_0)/Y_0$. The total social revenue is herbicide TUA revenues plus the value of the yield increase or:

$$SR_{it} = RTUA_{it} + P^*_t Y_t \left(\frac{Y_{it} - Y_0}{Y_{it}} \right) A_{it} \qquad (10)$$

where SR_{it} is the social revenue of firm i in year t; $RTUA_{it}$ is the TUA revenue of firm i in year t; P_t is the farm gate price of rape in year t; Y_t is the annual average weighted yield index of the Argentine varieties in year t; Y_{it} is the weighted yield index of the Argentine varieties of firm i in year t; Y_0 is the annual average weighted yield index of the Argentine varieties in year $t = 0$

(1960); and A_{it} is the area seeded of each variety of firm i in year t.

Notes

[i] The relative yield index of different canola varieties was converted to the same variety base (Torch) (1976 = 100). The yield index for each variety is obtained from the last reported value because it is thought to be a more accurate estimate of the actual yield performance.

[ii] Unfortunately, the whole series was not available from other provinces, so the data from Saskatchewan are used as a national proxy, which may be a reasonable approximation given that Saskatchewan is located in the centre of the oilseed rape-growing region.

[iii] The proportion of acres grown in Manitoba and Alberta was used to create the weighted average after 1990, which applied to the total oilseed rape acreage in the prairies (Statistics Canada, 2002).

[iv] The Canola Council data set was reduced by the average seed treatment costs for rape. The determination of the average seed treatment price for rape from 1991 to 2001 was based on David Blais (2003) (personal communication) and Jim Rogers (2003) (personal communication).

[v] Gaps in the data were filled in using the calculated annual average price of the seed per type per year and/or forecasting the seed price using the existing data as the underlying trend.

[vi] Monsanto charges a $37.6 TUA/ha fee for all Roundup Ready oilseed rape grown to extract value from producers. Two other companies that promote the development of herbicide-tolerant varieties, BASF (Clearfield oilseed rape sprayed with Odyssey herbicide) and Aventis (Liberty-Link rape sprayed with Liberty Herbicide), hold patents on the herbicides and can set the price wherever they want. The prices of Liberty and Odyssey herbicides are quite high when compared with Roundup. The calculated TUA per acre equivalents for these varieties are based on the notion that if BASF and Aventis faced a competitive market for their herbicides, they could be expected to sell their chemical for the price of Roundup, the excess revenue being rents. The herbicide costs used for this calculation was Roundup $20.13/ha, Odyssey $50.36/ha and Liberty $45.71/ha.

9 Mergers, Acquisitions and Flows of Agbiotech Intellectual Property

David Schimmelpfennig and John King

Economic Research Service (USDA), 1800 M Street NW, Rm 4179, Washington, DC, USA

Biotechnology has opened up new possibilities for research and commercial applications in agricultural markets. Innovations have originated in traditional agricultural input firms in the seed and chemical businesses, in entrepreneurial biotechnology start-ups and diversified multinational companies. The rush to pioneer new agricultural inputs has been intense, and many companies in the industry have experienced significant concurrent upheavals. As a result, a large number of firms with agricultural biotechnology (agbiotech) capability have been party either to a merger or to a corporate acquisition, or in several cases both.

A consequence of the numerous changes in ownership of agbiotech companies has been a dynamic element in knowledge ownership. This refers both to the ability of firms to acquire or possess knowledge and to their ability to protect knowledge with intellectual property rights. As increasing numbers of enterprises have been integrated through ownership changes, patterns of knowledge transfer have emerged. The overall trend in the industry towards consolidation as a result of mergers and acquisitions (M&As) is reflected in the consolidation of knowledge. However, this larger trend masks meaningful variation in the creation and transfer of knowledge.

To analyse significant aspects of knowledge transfer in agbiotech, we refer to a new assemblage of data linking firms with their intellectual property (IP). The Agricultural Biotechnology Intellectual Property (ABIP) database was developed through a collaboration between researchers at the United States Department of Agriculture (USDA) Economic Research Service and Rutgers University and can be accessed at www.ers.usda. gov/data/AgBiotechIP. This extensive data set tracks ownership histories of nearly 2000 public and private entities and over 11,000 US utility patents and other forms of IP affected by M&As between 1988 and 2002.

Many factors probably influence M&A activity in agricultural biotechnology. Market cycles, industry momentum and characteristics of the firms involved, including both their intellectual and tangible assets, may play a role in M&As. However, ownership and acquisition of knowledge motivate some M&A activity. Previous work on this topic has established some broad stylized facts. Graff *et al.* (2003) showed that firms tend to acquire complementary IP. Schimmelpfennig *et al.* (2003) added that acquisition of IP can be as significant a rationale for M&As as physical capital.

Defining Knowledge Flows

Measurement of knowledge is necessary to track changes in the possession of knowledge. Patents have several useful properties for such measurement. Patentable subject matter can extend to a wide variety of knowledge, encompassing abstract new techniques as easily as minor improvements to existing ones. Utility patents are an especially useful measure of knowledge because they represent knowledge in a standardized form. Although the invention described in each patent is presumably unique, the tightly structured format of each patent allows comparison and analysis across patents. For instance, patents contain application and grant dates that describe the emergence of knowledge through time. Patents typically include citations that indicate some of the provenance of knowledge contained in the patent. Also useful to researchers is the fact that patent claims describe the main significance of every patent. The new database relies on this feature for its classification of patents into the nine technology classes and 61 technology subclasses shown in Table 9.1.

Patents admittedly have some drawbacks as measures of knowledge. Strategic considerations affect the timing and extent to which patents disclose knowledge. Some knowledge – including some very important scientific advances – is not patentable because of prior research, restrictions on patent subject matter, or because of other interpretations of patent law. Also, many patented inventions fail to have much impact because they do not represent important knowledge, they become obsolete or they are not utilized for a variety of other reasons. Despite these disadvantages, patents represent knowledge that was deemed important enough by the applicant to incur the costs to protect with IP rights and by the Patent and Trademark Office to grant the patent. Possession of knowledge protected with IP can materially affect firm conduct and performance.

The patents used in this chapter are specifically well suited to measure the dynamic flow of knowledge between companies. First, the patent data indicate the initial assignee of IP rights in the patent at the time the patent was granted. Companion data on mergers and acquisitions from 1988 to 2002 allow an analysis of changes in patent ownership over time. Secondly, the patents were classified into technology classes and subclasses. The determinations concerning patent classification were based primarily on keyword searches of the full text of patents, but also made use of the US Patent Classification system and the International Patent Classification system. The patent classification system identifies heterogeneity in the types of patents most likely to change hands and ameliorates some of the pitfalls of using aggregate patent counts in empirical work. Finally, the data include other aspects of

patents such as citations and claims information that can serve as a measurement of knowledge impact or significance.

Handling Firm Heterogeneity

A previous International Consortium on Agricultural Biotechnology Research (ICABR) contribution by Klotz-Ingram *et al.* (2004) indicated that firms with different characteristics played different roles in the consolidation of agricultural input industries between 1985 and 2000. This chapter examines how different firm types influenced knowledge flows in agbiotech during this period. The categorization of firms into types is by no means independent of their IP; in fact, IP itself helps determine firm capabilities and therefore the category into which a firm fits. Fortunately it is possible to incorporate financial and business sector characteristics into IP ownership. The categories of firms are:

- **Agbiotech.** Mostly smaller US firms focused around biotechnology research and development capabilities.
- **Seed.** Firms with existing businesses supplying seed to farmers; an important source of cultivars and germplasm.
- **Chemical.** Firms primarily in the business of supplying agricultural chemicals such as fertilizer and herbicide; includes some companies with significant non-agricultural businesses.
- **Multinational/diversified.** Companies with several lines of agricultural business (seed, chemical, etc.) in several different regions.
- **European.** Arguably overlapping substantially with the previous category, but with a regional or management focus in Europe.

Firms were selected for our sample from the analysis of the financial characteristics of key firms done by Klotz-Ingram *et al.* (2004). All of the firms in that earlier work owned some IP and had publicly available financial information. After performing ownership analyses of these original firms using the new ABIP database, any firms that were either subsidiaries or parents of the first group were added to the sample. Finally, we included subsidiaries and other firms acquired by the ABIP firms. In all, the firms in our complete sample

Table 9.1. US utility patents for sample firms by technology class, 1976–2000.

Technology class	Subclasses	Patents in class
Plant technologies	Nutrition, agronomic applications, physical structure and plant function, plant organisms, cultivars, germplasm, male sterility/self-incompatibility, other plant technologies	1533
Patented organisms, non-plant	Transformed agricultural animals, cloned agricultural animals, other transgenic animals, microorganisms, fungi	269
Metabolic pathways and biological processes in plants	Plant growth regulators, enzymes, timing and control of gene expression	714
Metabolic pathways and biological processes in animals	Animal growth regulators, enzymes, timing and control of gene expression, animal reproduction	108
Protection, nutrition and biological control of plants and animals	Biotechnology pest regulation, microorganisms used for biological control, microorganisms used for other plant protection purposes, fertilizer, food and food additives from biotechnology	519
Pharmaceuticals	Veterinary pharmaceuticals, human pharmaceuticals, antibodies	161
Genetic transformation	Transformation platforms, mutagenesis, genetic markers, selectable marker techniques, culture growth, cell differentiation, transformation stability/heritability, diagnostic techniques	1068
Metabolic pathways and biological processes, DNA-scale	Recombination systems (CRELOX, FLP recombinase), RNA stability, protein stability, protein localization, transcription factor, transcriptional regulator, posttranscriptional regulator, receptor, T-DNA, Ds element, Ac element, Spm element, activation tagging, gene tagging, post-transcriptional modification	382
Genomics	DNA microarray, genome or EST sequencing, high throughput DNA sequencing, functional genomics, proteomics (including gene expression systems), protein sequencing, RDA, SAGE, subtractive hybridization, chromosome walking, chromosome landing, bulked segregant analysis, bioinformatics	78
Total		3003

The sum of patent classifications exceeds the total number of sample patents because approximately half (1504) of all patents are classified in multiple technologies.
Source: www.ers.usda.gov/data/AgBiotechIP

account for 3003 agbiotech patents issued between 1976 and 2002, approximately 27% of all agbiotech patents in the ABIP database. The numbers of patents in each technology class are listed in Table 9.1. Many patents in the database are held by US and foreign independent inventors, non-profit organizations and governments, and were not included in this analysis of knowledge flows among US and non-US companies.

Table 9.2 shows the relationships between the sample firm types. To illustrate consolidation in the agbiotech industry, Table 9.2 arranges the firms according to their original type (columns) and their type at the end of 2002 (rows). Firms along the

Table 9.2. Sample firms arranged by initial type and final type (based on merger/acquisition)

Final firm type	Initial firm type				
	Agbiotech	Seed	Chemical	Multinational	European
Agbiotech	Amylogene				
	Crop Genetics Intl				
	Exseed Genetics				
	Gene-trak				
	Rohto				
	Zoecon				
Seed		Delta & Pine Land			
		United Agriseeds			
Chemical	Agrigenetics	AH Robbins	American Cyanamid		
	Mycogen	Optimum Grains	American Home Products		
	Oncogene	Pioneer	Amoco		
	Visible Genetics		Bayer		
			Dupont		
			ICI		
			Knoll AG		
			Lubrizol		
			Merrell Dow		
			Miles Labs		
			Mobay Corp.		
			Nihon Tokushu		
			Quest Intl		
			Rohm		

Multinational

Agracetus
Calgene
DNA Plant Tech.

Asgrow
Dekalb
Holden's

Monsanto
Pharmacia
WR Grace

European

Mogen
Plant Genetic Systems

Garst
Northrup King

Advanta
Agrevo
Astra
Aventis
Ciba-Geigy
Hoechst
Institut Merieux
Merial
Novartis
Pepro
Purification Engineering
Rhone-Poulenc
Roussel
Sandoz
Schering-Plough
Zeneca

Source: www.ers.usda.gov/data/AgBiotechIP

Table 9.3. Firm financial data by firm type.

	Five-year average for firms in sample by type			
	R&D intensity (expenditures/sales)	Gross profit margin	Debt-to-equity	Assets-to-equity
Agbiotech				
1985–1989	0.88	−35.33	11.21	1.16
1990–1995	1.01	32.61	14.74	1.31
1995–2000	0.24	13.73	25.77	1.89
Seed				
1985–1989	0.09	52.66	15.12	1.71
1990–1995	0.09	54.43	22.52	1.73
1995–2000	0.1	50.97	42.23	2.19
Chemical				
1985–1989	0.06	40.06	46.51	2.18
1990–1995	0.08	39.8	54.51	2.6
1995–2000	0.06	31.67	78.25	2.98
Multinational/European				
1985–1989	0.06	43.21	83.43	2.7
1990–1995	0.07	45.41	83.93	2.98
1995–2000	0.1	51.16	362.7	7.99

Source: Klotz-Ingram *et al.* (2004).

diagonal did not change type, while firms off the diagonal were acquired by a firm of a different type. This illustration understates the extent of consolidation because several firms merged with or were acquired by other firms of the same type.

Table 9.3 provides some firm financial data indicating that the characteristics of the firms in the different categories were quite different and that their financial positions changed dramatically between 1985 and 2000. The 5-year averages in the table show that R&D intensity (calculated as the proportion of R&D expenditures to sales) was much higher for the agbiotech companies and that by the 1995–2000 period, intensity had fallen as industry consolidation was taking place. Gross profit margin (GPM) measured as net sales minus the cost of goods sold (converted to a percentage) represents the markup on goods sold and is an indicator of the aggregate pricing strength of the companies in a category. Here again, the agbiotechs were in upheaval and their average

GPM goes from negative to strongly positive and back down, but even at the peak is below any of the other company categories during the same period.

Also revealing for M&A activity is that the multinationals started with the highest debt-to-equity ratios and ended with an average ratio that was 14 times that of the agbiotechs. Assets-to-equity were similarly high for the multinationals in the last period. This seems to indicate that the multinationals were extending themselves beyond what the more conservative chemical companies were willing or able to manage in 1995–2000 across all their lines of business.

Measuring Knowledge Flow

One of the most remarkable aspects of the agbiotech knowledge possessed by these companies is its high degree of fluidity between company

Table 9.4.   Flows of agricultural biotechnology patents.

Final firm type	Initial firm type				
	Agbiotech	Seed	Chemical	Multinational	European
Agbiotech	24 (5%)				
Seed		31 (5%)			
Chemical	219 (49%)	451 (69%)	651 (100%)		
Multinational	175 (39%)	175 (27%)		528 (100%)	
European	31 (7%)				718 (100%)
Total	449	657	651	528	718

Percentages in parentheses indicated the proportion of patents by initial firm types. Columns may not sum to 100% due to rounding. Blank cells indicate zero.
Source: www.ers.usda.gov/data/AgBiotechIP

types. The off-diagonal patent counts in Table 9.4, added together, are the patents which flowed from one category of firm to another. This number divided by the total represents 35% of the patents in the sample. This figure understates somewhat the extent of technology flows for two reasons. First, it ignores knowledge flows within firm types. Secondly, patents issued to subsidiaries subsequent to their acquisition by firms of another type are counted as belonging to the parent. For instance, agbiotech firm Agracetus was granted 13 patents prior to its acquisition by the diversified multinational firm Monsanto in 1996, and received another ten as a Monsanto subsidiary. Likewise, seed company DeKalb Genetics had 33 agbiotech patents prior to its 1998 acquisition by Monsanto, but 79 subsequent patents naming DeKalb as the assignee are attributed to Monsanto.

Table 9.4 also reveals that most (95%) agbiotech and seed company patents contributed to knowledge flows across firm types, while no chemical, multinational and European company-owned patents in the sample flowed across firm types. The simple explanation is that many of these agbiotech and seed companies were acquired by firms of a different type and the other companies were not. Notwithstanding, this perspective illustrates a tremendous disparity in the extent and direction of knowledge flows among the different firm types.

Adding in changes in patent ownership within firm type raises the overall percentages as would be expected. However, over half (65%) of the patents in our sample changed hands between original assignee and the owner in 2002. Extending the flow of knowledge as a river or stream analogy used thus far, the rate of patent 'churn' (as in an eddy current) within company categories is most pronounced for the agbiotech and seed companies who had 94% of their patents change hands in addition to contributing almost all of the patents flowing between company categories. Since some of the largest knowledge creators and holders of patents (Dupont, Dow and Bayer among the chemical companies for instance) did not change hands from 1988 to 2002, this rate of churn is quite remarkable.

Knowledge flows across firm types also vary with technology classification. Based on the type of the initial assignee, agbiotech and seed companies generated 54 and 60% of plant technologies and genetic transformation patents, respectively. By the end of the sample period, chemical and multinational firms controlled 76% of plant technology patents and 79% of genetic transformation patents.

Analysis of technology flows in different technology classes in Table 9.5 provides a more detailed picture of the sources, acquisition and changing patterns of knowledge in this industry. For instance, seed companies have been perhaps the most copious source of knowledge in plant technologies and genetic transformation. This is consistent with seed company knowledge in plant breeding and ownership of large stocks of germplasm. Knowledge in the

Table 9.5. Knowledge flows across firm types by technology class.

Technology	Final firm type	Initial firm type				
		Agbiotech	Seed	Chemical	Multinational	European
Plant technologies	Agbiotech	14				
	Seed					
	Chemical	164	31	126		
	Multinational	63	374		263	
	European	16	170			312
Patented organisms, non-plant	Agbiotech	1				
	Seed					
	Chemical	33	6	65		
	Multinational	25			43	
	European	3				93
Metabolic pathways/biological processes in plants	Agbiotech	9				
	Seed					
	Chemical	17	2	226		
	Multinational	57	29		134	
	European	6	17			217
Metabolic pathways/biological processes in animals	Agbiotech					
	Seed					
	Chemical		3	49		
	Multinational				36	
	European					20
Protection, nutrition and biological control in plants and animals	Agbiotech	12				
	Seed					
	Chemical	168	16	90		
	Multinational	11	2		62	
	European	3				155

Category	Firm type	Agbiotech	Seed	Chemical	Multinational	European
Pharmaceuticals	Agbiotech	4				
	Seed					
	Chemical	5		82	32	37
	Multinational	1				
	European					
Genetic transformation	Agbiotech	8	14			
	Seed	47				
	Chemical	80	325	142	93	193
	Multinational	9	157			
	European					
Metabolic pathways/biological processes, DNA-scale	Agbiotech	31	6			
	Seed	60				
	Chemical	14	51	83	35	99
	Multinational		3			
	European					
Genomics	Agbiotech	7	4			
	Seed		13			
	Chemical		2	18	12	22
	Multinational					
	European					

Cell contents indicate the number of US agbiotech utility patents.
Source: www.ers.usda.gov/data/AgBiotechIP

two classes of plant protection, nutrition and biological control patents was largely generated by agbiotech companies; this suggests that entrepreneurial start-up firms were focused initially on technology applications to increase yields or to substitute for agricultural inputs. Chemical firms and European firms led the creation of knowledge in metabolic pathways and biological process areas as well as pharmaceuticals (related to agricultural biotechnology). These technology classes include plant and animal growth regulators and the synthesis of fine chemicals, which are related to competencies possessed by these firms in the manufacture of inputs such as pesticides and herbicides.

The tendency of knowledge to remain within a single company type is equally present in the individual technology types of Table 9.5 as it is in Table 9.4. However, some technologies flowed across company types more readily. Seed company knowledge in plant technologies (~33% of the total) flowed primarily to chemical companies, largely a result of the acquisition of Pioneer Hi-Bred by Dupont. Seed company knowledge in plant technologies also flowed to multinational/diversified companies through the acquisitions of Holden's Foundation Seed and DeKalb Genetics by Monsanto.

Knowledge flows in transformation technologies appear similar in size and proportion to knowledge flows in plant technologies, mostly as a result of the same mergers. However, genetic transformation knowledge flows involved a greater number of firms within each type. This might indicate the central role of transformation technologies in this industry. More firms acquired knowledge in this area, and firms of the larger chemical, multinational and European types acquired knowledge from a greater variety of sources within their own types (churn) and among agbiotech and seed companies (flow).

The pattern in both Table 9.4 and Table 9.5 that knowledge produced by firms in the larger types (chemical, multinational and European) tends to remain there is reasonably consistent across technology classes. Although it is unsurprising that larger firms acquired smaller seed and agbiotech firms rather than the other way around, it is not altogether apparent why knowledge would not flow between firms among the larger types. Blockbuster multi-billion dollar company acquisitions were not as common in this industry as they have been in some others. A similar observation is

that, with a few important exceptions, knowledge flows were not trans-Atlantic. European companies did not acquire the smaller companies in the seed and agbiotech categories, instead acquiring other European firms to obtain the corresponding knowledge flows. This might explain some of Novartis' motivation for making a US$25 million plant research deal in November 1998 with the University of California, Berkeley (Dalton, 1999). The absence of non-US patents to measure knowledge flows is probably a factor here as well, to the extent that owners of technology file for different patents in the USA and Europe.

Table 9.6 summarizes knowledge creation by technology class to describe areas of emphasis among the different firm types. To summarize the degree to which firms in a given type specialize within a technology, we constructed a statistic to measure technological 'focus'. Based on the Herfindahl–Hirschmann Index (HHI), this statistic is calcualted as $\Sigma_k s_k^2$, where s_k indicates the share of patents in technology k expressed as a decimal. Unlike the HHI, this statistic can exceed unity because the patent classifications are not mutually exclusive. To adjust for the quadratic effect of patents with multiple classifications, we also report the statistic $\Sigma_k(s_k^2/\bar{n})$, where $\bar{n}$ is the average number of classifications for each patent within a given type.

The two statistics produce parallel results: seed companies and agbiotech companies were the 'most focused' (least diversified), chemical companies were the 'least focused' (most diversified), and the multinational and European companies were in between (but closer to the chemical companies). Note that the larger companies in the chemical, multinational and European categories are not diversified directly as a result of their investments in other lines of business: the patents in the database were selected to include only those related to agbiotech. The widespread knowledge and patent holdings of these larger companies is well known and not reflected in the database.

The emphases of the seed companies were in plant technologies and genetic transformation, while chemical, multinational and European companies each had around 30% of their patents in transformations and 40–55% in plant technologies. The larger firms had quite similar portfolios of patents in the other classes as well, with 25–30% in metabolic pathways, and 10–20% in protection, nutrition. In fact, the main difference

Table 9.6. Technology emphasis and focus by firm type.

	Agbiotech (236 patents)	Seed (344 patents)	Chemical (907 patents)	Multinational (711 patents)	European (710 patents)
Plant technologies	61%	88%	38%	55%	44%
Patented organisms, non-plant	14%	1%	8%	7%	13%
Metabolic pathways and biological processes in plants	16%	8%	27%	25%	31%
Metabolic pathways and biological processes in animals	0%	1%	5%	5%	2%
Protection, nutrition, and biological control of plants and animals	53%	4%	16%	10%	21%
Pharmaceuticals	4%	0%	8%	4%	4%
Genetic transformation	28%	70%	34%	32%	27%
Metabolic pathways and biological processes, DNA-scale	19%	5%	14%	9%	14%
Genomics	<1%	3%	2%	25%	3%
Technology 'focus'[a]	0.81	1.28	0.39	0.56	0.45
Classification-adjusted technology 'focus'[a]	0.30	0.50	0.20	0.24	0.23

[a]See text for description of technology 'focus' measurements.
Source: www.ers.usda.gov/data/AgBiotechIP

between the firms in these categories is that only the multinationals have a large share of genomics patents.

Quality (in Addition to Quantity) of Classified Patent Flows

The number of times a patent is cited by another patent within the patent sample as well as outside in other utility patents gives some indication of the quality of the patents being acquired. In fact, patents held by agbiotech firms are cited more than the patents held by all the other categories of firms combined (for patents within the database), providing a quality motivation for the quantity flows seen above. Seed companies held the second most cited group of patents (within the database) but the lowest cited patents by all utility patents. The number

of patents they cite (within the database) is actually higher for seed company patents than agbiotech patents. As might be expected, patents that appear in more than one technology classification tend to be the ones cited more often.

The citation rates for patents that changed hands by technology classification indicate how knowledge quality flows differ between technologies. With all of the M&A acquisition activity in this industry, it might have been expected that firms were trying to acquire the highest quality patents, but citations for patents that did not change hands were higher than for those that did for six out of nine technology classifications (citations within the database). The six include: patented organisms, metabolic pathways and biological processes in animals, pharmaceuticals and genomics, in addition to the pivotal plant technology and genetic transformation patents.

Table 9.7. Flows of patent quality by technology class.

Technology	Citations received by patents with same assignee and final holder		Citations received by patents that changed hands	
	Mean for agbiotech patents	Mean for all patents	Mean for agbiotech patents	Mean for all patents
Plant technologies	1.76	2.29	1.62	1.78
Patented organisms, non-plant	2.25	3.44	1.39	1.87
Metabolic pathways and biological processes in plants	0.89	2.97	1.44	2.64
Metabolic pathways and biological processes in animals	1.37	4.44	0.08	1.18
Protection, nutrition and biological control of plants and animals	2.12	3.88	2.43	3.09
Pharmaceuticals	1.73	5.48	1.03	2.31
Genetic transformation	1.81	3.57	1.09	1.36
Metabolic pathways and biological processes, DNA-scale	1.59	2.37	1.94	2.23
Genomics	1.58	2.71	0.44	0.78

Source: www.ers.usda.gov/data/AgBiotechIP

When considering citations of all patents (not just those in the agbiotech database), the mean citation rate is higher in all nine classifications for patents that remained with their original holders than those that changed hands. Rather than indicating that firms were acquiring less-cited (lower quality) patents, this could mean that either the firms making acquisitions already held many of the highest quality patents or some holders of high quality patents successfully resisted M&As.

Conclusions

This chapter provides some preliminary evidence that knowledge flows, represented in different technology classifications of US Patent and Trademark Office agbiotech utility patents, were substantial and substantially different between different types of firms in the agricultural inputs industry. The financial data indicate that the chemical companies enjoyed more stable selected financial ratios across their lines of business between 1985 and

2000, while acquiring agbiotech and seed company plant technology patents and genetic transformation technologies at over twice the rate of the multinationals with only one exception. The multinationals acquired more transformation patents from the agbiotechs than did the chemical companies. The quality of patents (by citations) indicates that the highest quality patents were not even changing hands and were presumably developed in-house by the chemical, multinational and European firms.

The data used in this chapter have been 3 years in development, and are just beginning to be explored. The database is publicly available and web-searchable and offers great potential for empirical work. The USDA/CSREES funded the work through the Initiative for Future Agriculture and Food Systems (IFAFS) and any comments or suggestions would be appreciated. The IFAFS project has produced outputs in addition to the database, and the project's website can be accessed at http://aesop.rutgers.edu~agecon/ifafs/ The USDA provided generous support for much of this work through the grant as well as staff and facilities at the Economic Research Service, but the views expressed are not necessarily those of the USDA.

References

Dalton, R. (1999) Berkeley dispute festers over biotech deal. *Nature* 399, 5.

Graff, G.D., Rausser, G.C. and Small, A.A. (2003) Agricultural biotechnology's complementary intellectual assets. *Review of Economics and Statistics* 85, 349–363.

Klotz-Ingram, C., Schimmelpfennig, D., Naseem, A., King, J. and Pray, C. (2004) How firm characteristics influence innovative activity in agricultural biotechnology. In: Evenson, R.E. and Santeniello, V. (eds) *The Regulation of Agricultural Biotechnology.* ICABR, CAB International, Wallingford, UK.

Schimmelpfennig, D., King, J. and Naseem, A. (2003) Intellectual capital in a Q-theory of agbiotech mergers. *American Journal of Agricultural Economics* 85, 1275–1282.

10 The Impact of Regulation on the Development of New Products in the Food Industry

Klaus Menrad[1] and Knut Blind[2]

[1]*University of Applied Sciences of Weihenstephan, Science Centre Straubing, Schulgasse 18, D-94315 Straubing;* [2]*Fraunhofer Institute for Systems and Innovation Research (ISI), Breslauer Str. 48, D-76139 Karlsruhe, Germany*

The question of regulation, innovation and their impact on competitiveness in global markets is of high relevance for all industries. However, little has been done to understand the effect of regulation on the capacity of a traditional industry such as the food industry to innovate and introduce new products and services in the market. The debate has taken place at a level of anecdotal evidence and poor systematic empirical foundations. In addition, most of the approaches assume a static framework, not recognizing the long-term dynamic feedback loops between regulation and technical progress and new markets. Finally, the majority of the expressed statements, especially from industry, come to the conclusion that the negative impacts of regulation outweigh the positive effects. This chapter aims to bridge the gap between the challenge to shape a regulatory framework, which allows the emergence of new markets, and even to use regulation as an instrument to foster innovation and the lack of adequate, reliable and systematic knowledge on their inter-relationship.

For the study, a methodological set was used, consisting of reviewing existing literature and reports, marketing statistics, press releases and other documents of companies and trade associations as well as regulatory and legal documents. In addition, a questionnaire-based survey among manufacturing companies and research institutions of the European Union (EU) was carried out in order to analyse the view of different industries – among others the food processing industry – concerning the regulatory framework in the EU and its inter-relationships to innovation activities in different fields. Finally, expert interviews were used to complement the picture and give additional background insight into the analysed processes.

Regulatory Regime of the Food Industry

Due to the increasing internationalization of food and commodity markets, political and regulatory influential factors gain increasing relevance for the food industry. By speeding up the integration of international markets and increasing numbers of international joint ventures, the food industry is more and more influenced by international legislation. Of specific relevance for food production and food processing are the standards of the so-called *Codex Alimentarius* which comprises all standards, voluntary agreements and recommendations of the so-called *Codex Alimentarius* Commission. This Commission is the highest international committee for defining worldwide accepted standards for foods. In 2000, membership of the *Codex Alimentarius* Commission comprised 165 countries representing 98% of the world's population. In 2003, the *Codex Alimentarius* contained more than 230 standards, more than 3000 upper levels of pesticide residues and more than 1000 assessments of food additives (*Codex Alimentarius*, 2003). The Codex develops standards or gives recommendations for labelling issues, food additives, dietary food products, harmful substances in food, analytical methods, aspects of general food hygiene, the control of food imports and exports, as well as levels of residues of veterinary pharmaceuticals and pesticides in foods. The high relevance of this type of standard is underlined by the fact that an increasing number of countries are transforming the Codex's standards into national law.

The EU legislation on foodstuffs (with some exceptions such as novel foods and novel food ingredients) leaves food industry companies free to market their products without pre-market approval. Food manufacturers have to ensure that their products are safe and do not mislead the consumer. These requirements must be met under the sole responsibility of the company and are subject to post-marketing controls by public authorities. In April 1997, the European Commission published the Green Paper *The General Principles of Food Law in the European Union* which defines a regulatory framework which covers the entire food chain. This document had the objectives to ensure a high level of protection of public health and safety and of consumer protection, to ensure the free movement of goods within the single market, to base legislation on scientific evidence and risk assessment, to ensure the competitiveness of the European industry and enhance export prospects, to place the primary responsibility for safe food with industry, producers and suppliers, as well as to ensure that legislation is consistent, rational and clear (European Commission, 1997).

Following a series of food scandals in the 1990s, the European Commission suggested in a White Paper on Food Safety in January 2000 the establishment of an independent European Food Safety Authority (EFSA) which should be responsible for independent scientific advice on all aspects related to food safety, operation of rapid alert systems and communication of risks (European Commission, 2000). In addition, suggestions for a new legal framework concerning food safety, control activities, consumer information and international arrangements related to food safety were given in the White Paper. Furthermore, the European

Commission announced its intention to streamline and simplify the EU decision-making process for foodstuffs 'in order to ensure efficacy, transparency and rapidity' (European Commission, 2000).

In the USA, foods are regulated under the Federal Food, Drug and Cosmetics Act (FFDCA). This Act was the first and most comprehensive law in the world covering production, distribution and trade of foods, drugs, medical devices and cosmetics. The FFDCA defines foods and standards for food, adulteration of food as well as regulations for misbranding. Under FFDCA, the Food and Drug Administration (FDA) oversees safety and labelling of food products, with the exception of those containing meat or poultry which are controlled by the United States Department of Agriculture's Food Safety and Inspection Service (FSIS).

In the USA, there are no federal requirements that food manufacturing companies have to be registered or obtain pre-market approval of food. However, for some categories of food, such as seafood and low-acid canned foods, complex quality control programmes for manufacturing are mandatory (Greenberg, 2000). The US food safety system is based on strong but flexible science-based legislation and industry leaders' responsibility to produce safe foods. The regulatory approach used in the USA to control food safety refers to a so-called Hazard Analysis and Critical Control Points (HACCP) system (Greenberg, 2000). The US food safety programmes are risk-based in order to ensure that consumers are protected from health risks of unsafe food. Decisions within these programmes are mainly science-based and involve risk analysis processes. Regulations in the food safety area are often developed and revised in a public process that encourages participation by the regulated industry, consumer organizations and other stakeholders throughout the development and promogation of a regulation.

Another important aspect of food regulation in the USA is labelling requirements of foods. The two most relevant legislations in this area are the Fair Packaging and Labeling Act (FPLA) and the Nutrition Labeling and Education Act (NLEA) of 1990 which is an amendment to the FFDCA. Before the enforcement of the NLEA, the FDA ran a voluntary nutrition labelling programme which was in place for more than 20 years (Storlie and Brody, 2000). In 1990, the US Congress passed the NLEA which made nutrition labelling mandatory on most packaged food products and mandated that FDA initiate uniformity in the content and format of the nutrition label on the package. In addition, the NLEA set up the regulatory framework for the approval of health claims which are of specific relevance for the development and marketing of functional food.

Impact of Regulation on Innovation Activities in the Food Industry

In the following, three very innovative fields in the food industry are analysed in terms of whether the existing regulatory framework has hindering or facilitating impacts on the development and introduction of new products: the use of genetic engineering for food production and food processing; the field of health-oriented functional foods; and organic food products.

Genetically modified organisms (GMOs) and novel foods

Since the mid-1990s, genetically modified (GM) plants have been marketed and cultivated which directly, or via animal feed, can enter the food chain. By their protagonists, genetic engineering approaches are regarded as major tools to increase productivity and efficiency in food processing in the future (Garza and Stover, 2003). On the other hand, an intensive public debate is carried out globally concerning the safety of these approaches and derived novel foods, potential harm for the environment and human or animal health, as well as their socio-economic impacts (Otsuka, 2003). Since GMOs and derived novel food products represent new developments in the area of food production and food processing, there have been relatively restricted experiences with this type of product. Therefore, state authorities took specific actions to deal with the potential risks of GMOs. The general targets of the respective legislations are to ensure human health when consuming GMOs or derived novel foods, to prevent or minimize the potential harm of GMOs to the environment as well as to provide the necessary information in order to ensure the freedom of choice of consumers or users of such products. In particular, the EU policy related to GMOs was intensively influenced by the emergence of bovine spongiform

encephalopathy (BSE) and other food crises during the 1990s, public criticism and undermined trust in public authorities to manage such crises adequately as well as the low consumer acceptance of agro-food biotechnology (Gaskell *et al.*, 2000; Eurobarometer, 2001; Loureiro, 2003).

The fundamental question which arises concerning regulation of GMOs is whether GM crops or other GMOs have to be acknowledged like conventional crops or organisms, and therefore it is sufficient to use the general legislation valid for such crops or organisms or whether it is necessary to adopt different and specific regulations for GMOs. In the USA, GM crops are considered specific and different in terms of intellectual property rights since a patent can be granted to them but not to conventional crops. On the other hand, the introduction of GM crops in the environment and into the market follows the principle of 'substantial equivalence' and therefore the same steps are required as for conventional crops (Esposti and Sorrentino, 2002).

The EU takes the opposite approach concerning regulation of GMOs compared with the USA. Even after Directive 98/44//UEC, patents cannot be granted to GM crops, but they are protected by the same breeders' rights acknowledged to conventional crops. In contrast to the US procedure, the EU approach for environmental release and market approval of GMOs follows a rather strict interpretation of the 'precautionary principle'. For this purpose, specific regulations have been put into force dealing with GMOs which require different and often more complex procedures than for conventional products (Esposti and Sorrentino, 2002).

GMOs have been regulated by the EU since the beginning of the 1990s. The EU Directives 90/219/EEC and 90/220/EEC were the first regulations which tried to establish a system for controlling R&D and commercialization of GMOs in the EU. These regulations were designed to protect citizens' health and the environment, and addressed authorization, labelling and traceability issues relevant for GMOs. Since its enpassment in the year 1990, Directive 90/220/EEC was criticized by different stakeholder groups. In addition, all notifications for market approval of agricultural GMOs raised concerns of one or several EU member states during the 1990s (Sauter and Meyer, 2000). Therefore, in June 1999, a *de facto* moratorium on commercialization of GMOs was agreed by the Community's Council of Environmental Ministers to suspend all approval applications for GMOs until implementation of the revised Directive 90/220/EEC, in order to provide a stricter legal framework covering not only safety issues but also labelling and traceability of GMOs (Lheureux *et al.*, 2003).

Directive 2001/18/EC on the deliberate release into the environment of GMOs was passed in February 2001 and replaced Directive 90/220/EEC. This Directive modified the rules for environmental release and market approval of GMOs significantly by restricting market approval to 10 years and in terms of the requirement for post-market monitoring of each GMO. In Regulation 1829/2003/EC, a tolerance level of 0.9% is foreseen for adventitious admixture of GM material: food products which exceed this level have to be labelled accordingly. These labelling rules for GM-derived novel foods come into force irrespective of whether DNA or protein of GM origin can be found in the final product. Another area of intensive debate is to ensure coexistence between GM crops, conventional and organic farming as well as the question of liability in case economic damage results from admixture of gene flow (European Commission, 2003a).

In contrast to the EU, in 1992 the US FDA outlined a policy that did not require the market approval for GM crops and placed the responsibility for investigating and reporting potential problems associated with GM foods on the companies. During this phase, the FDA recommended voluntary labelling of specific GM foods but opposed their mandatory labelling (Food and Drug Administration, 1992). The practice of the FDA was modified in 2001 when the Authority announced a mandatory notification by manufacturers for plant-derived bioengineered foods (Food and Drug Administration, 2001).

In the EU, there is still a broad pipeline of R&D activities related to agricultural and food GMOs which is fuelled by differing organizations such as large multinational companies, SMEs (small and medium sized enterprises, i.e. < 500 employees), universities and non-university research institutions (Lheureux *et al.*, 2003). However, EU publication intensity in plant biotechnology grew significantly less than average compared with other fields of biotechnology (Reiss and Dominguez Lacasa, 2003). In addition, field trials with GMOs have dropped by 76% since the introduction of the

de facto moratorium (Lheureux *et al.*, 2003) and more than half of the companies mentioned in a survey carried out in 2002 had cancelled R&D projects related to GMOs in recent years. Companies highlighted the unclear legal situation in the EU, high costs and time requirements for safety testing, low consumer and user acceptance of GM products as well as uncertain future market perspectives as the main obstacles for GMO-related innovation activities in the EU (Lheureux *et al.*, 2003). Although 14 GM plants received market approval from the EU before 1998 (and 20 are awaiting market approval), only 32,000 ha of Bt maize were commercially planted in Spain in 2003 (James, 2003). Altogether, the impression is that the *de facto* moratorium has had a negative impact on pre-market innovation activities related to GMOs in the EU (Menrad, 2003b). This relates in particular to SMEs which often have given up such projects due to their limited financial and personnel resources.

In contrast to the situation in the EU, field trials with GMOs have not fallen significantly in the USA in the last 6 years. In addition, a high number of mainly herbicide- and/or insect-resistant GM plants have been approved for commercial use in the USA (Agriculture and Biotechnology Strategies, 2004). With almost 43 Mha in 2003, the USA also has the worldwide leadership concerning commercial cultivation of GM plants (mainly soybeans, maize and cotton) (James, 2003), while in Canada particularly GM rapeseed is grown for commercial use.

The EU regulatory framework adopted during the 1990s has played an important, largely negative role in the development of GMOs in the EU in the last decade. During this time period, increased regulatory oversight in agro-food biotechnology coincided with growing negative public opinion and diminished trust in public authorities and regulatory agencies. In this context, companies regarded the 'constantly changing regulatory environment' as one major constraint for R&D and commercialization of GMOs in the EU. In particular, the practical handling of the existing regulations was strongly criticized as being too slow, bureaucratic and causing extraordinary costs. Politics was criticized for not taking any clear decision regards GMOs (which will form a reliable planning basis for the companies) and periodically intervening in the regulatory processes. Combining the findings of the different studies (Lheureux *et al.*, 2003; Reiss and Dominguez Lacasa, 2003) provides evidence that the unclear legal situation with respect to the commercialization of GMOs led to cutting down research activities in plant biotechnology in the EU which can be measured as decreasing scientific output. In more general terms, the unclear legal situation related to GMO on the commercial side seems to have a negative feedback on the science base.

Functional food

In addition to satisfying hunger and providing humans with the necessary nutrients, functional food is intended to provide additional benefits to consumers by preventing nutrition-related diseases and increasing the physical and mental well-being of humans. For this purpose, functional food contains specific 'functional' ingredients to which are attributed particular effects on human health or well-being. These substances aim at reducing the risk of, for example, coronary heart disease, specific types of cancer, diabetes or osteoporosis, which are amongst the most important causes of death in many industrialized countries (Menrad et al., 2000).

From a legal point of view, functional food is positioned in a transitional zone between food and pharmaceuticals. In the EU, its member states and the USA, food and pharmaceutical products are subject to different regulation regimes and are – at least in the EU – regulated by differing relevant authorities. Another fundamental difference between these two product fields is that pharmaceuticals and other medicinal products require pre-market approval by state authorities, while EU and US legislation on foodstuffs (with a few exceptions such as, for example, novel foods or food additives) leaves companies free to market their products, if they are safe and do not mislead consumers, and afterwards they are subject to post-marketing control by public authorities. Since functional food is positioned in a 'grey zone' between these two areas, high uncertainty emerges both for commercial activities of companies and for consumers. Due to the novelty character of functional food, additional questions arise concerning the safety and efficacy of such products, the required testing and monitoring methods, their impact on consumers' nutritional behaviour and the institutional procedures and responsible authorities (Menrad et al., 2000).

In the EU, there exists no harmonized or specific regulatory framework nor a precise legal definition of functional food, i.e. the general regulations valid for food are also relevant for functional food. However, there are differences in the practical handling of these general regulations with regard to functional food between the EU member states (Groeneveld, 2000) – which hinder the development of a common market for such products. Due to the absence of specific regulations, the so-called Novel Food Regulation (Regulation 258/97/EC) was used for single functional food products applying for market approval in the EU in recent years. Another important bottleneck for functional food represents the fact that there exists no harmonized legislation on health claims at the Community level so far, which results in differences in the practical handling of such claims between the member states (Menrad, 2003a). Under the existing regulatory framework in the EU, it is prohibited to attribute to any foodstuff the property of preventing, treating or curing a human disease or referring to such properties. The regulation on nutrition and health claims proposed by the European Commission in July 2003 includes far-reaching suggestions concerning the use and registration requirements of generic claims as well as for the requirements to approve specific health-related claims. In addition, it is foreseen that it will prohibit certain claims related to, for example, general well-being or slimming (European Commission, 2003b).

In contrast to the EU situation, food manufacturers in the USA can use about 13 generic claims to inform consumers about the health benefits of functional food products which have been approved by the FDA since 1990 (Ringel Heller and Silverglade, 2001). The regulatory framework for these activities was provided with the passing of the NLEA in 1990, which set up the regulatory framework for the approval of health claims. In order to protect consumers from unproven health claims, the respective claims have to be supported by the totality of publicly available scientific evidence and there must be 'significant scientific agreement' among qualified experts (Heasman and Mellentin, 2001).

Despite definition problems and highly varying market figures for functional food, it can be concluded that the market for functional food has developed faster and to larger market values in the USA compared with the EU. While the total

US market for functional food is estimated as US$15–20 billion, which equals around 2% of the US food market (Centrale Marketing-Gesellschaft der Deutschen Agrarwirtschaft mbH, 2002; Marra, 2003), the corresponding EU sales figures amount to around €4–8 billion, which equals a functional food market share of below 1% of the total EU food and drinks market. Within the EU, Germany, France, the UK and the Netherlands represent the most important markets for functional food, while Mediterranean consumers showed limited interest in this type of food in recent years (Hilliam, 2000; Menrad, 2003a).

According to the available future market estimations, it can be assumed that functional food will increase its market volume considerably in the coming years. Most market estimations assume that 3–5% of the food market represents the growth limit for functional food in Europe in the coming 10 years (Dustmann and Weindlmaier, 2002; Menrad, 2003a). In this sense, functional food has a significant growth potential but it represents a multiniche market with a high number of limited product segments and very few high-value product categories. The most important drivers of innovations in the field of functional food are scientific and technical developments in nutrition-related research, the interest of food companies in participating in growing segments of an almost stagnant food market as well as consumers' interest in innovative products supporting their health and wellbeing. So far, multinational food companies as well as internationally acting food ingredient suppliers are best positioned to overcome specific challenges during the R&D and marketing process of functional food, while small and medium-sized food industry companies only have limited opportunities in this field (Menrad, 2003a).

A major bottleneck for the future development of functional food in the EU represents the unclear legal situation concerning procedure and factual requirements for market approval of such products as well as the use of health claims. This lack of a clear regulation and definition of functional food and non-harmonized procedures between the EU member states impedes the potential growth of this market. Thus, the EU food industry cannot take full advantage of a potential growth segment in the mature and stagnating food market in the EU. In this sense, an important basic activity with a mid-term perspective is the clarification and standardization of the regulatory status

of functional food in the EU both for consumers and industrial companies as a legal basis for economic activities (Menrad, 2001).

Organic agriculture and food production

At the end of the 1980s, both the EU and USA took initiatives to develop a specific regulatory framework for production, processing and trade of organic products. Major targets of the related policies were reducing costs of market stabilization measures for conventionally produced agricultural commodities, supporting environmentally friendly ways of agricultural production and protecting consumers from being misled in the area of organic foods (Lampkin *et al.*, 1999). While the US policy focused almost solely on defining standards for producing, processing, marketing and importing organic products (Greene and Kremen, 2003), the European Commission followed a dual strategy which first supported conversion of conventional farms to organic agriculture by direct payment schemes and advisory assistance, and secondly improved market transparency by defining, implementing and controlling clear standards for production, processing and marketing of organic products (Lampkin *et al.*, 1999).

Regarding timing of the respective legislation, the USA took the lead when passing the Organic Food Production Act (OFPA) in 1990 by the US Congress. However, it took more than 10 years until the OFPA was fully implemented and nationwide standards for organic products were put into force in the USA (Greene and Kremen, 2003). With Regulation 2078/92/EC, a common regulatory framework was established for EU member states to implement policies to support organic farming which was introduced in most countries in 1994 and 1995. An important part of this Regulation is direct payments for conversion to organic farming for which maximum rates eligible for co-financing by the EU are defined in the Regulation. Another key element of EU legislation in the field of organic food products is Regulation 2092/91/EEC which created a common legal definition for organic farming and the respective products. This Regulation required that all fresh and processed products of plant origin must meet the organic standards defined in the Regulation and that producers and operators must submit their business activities to a publicly controlled inspection system. In 1999, analogous requirements were put into force for production and marketing of organic livestock products (Lampkin *et al.*, 1999).

Both in the USA and in the EU, the introduction of the organic food regulation has had a strong impetus for supply and demand of such products. However, EU farmers and food processors reacted faster and in much higher numbers in order to benefit from the market opportunities provided by organic foods. This can be illustrated by the fact that by 2000, around 2.1% of all farms which operated 3.1% of the total utilized agricultural area of the EU (Organic Centre Wales, 2003b) had converted to organic agriculure in comparison with 0.32% of all US farms being organic in 2001 with 0.25% of the operated farm area (United States Department of Agriculture, 2003; Yussefi and Willer, 2003). In particular, producers of plant products such as fruits and vegetables, potatoes or cereals, increased organic production significantly during the 1990s, but, at least in the EU, organic livestock production (e.g. milk and eggs) seems to have grown at high rates in recent years (Hamm *et al.*, 2002).

Besides higher price premiums for organic products, this much stronger increase of domestic organic production in the EU can be linked to the direct payments provided by the European Commission and national governments in order to support conversion to organic agriculture. Since conversion of conventional farms to organic agriculture reduces natural yields significantly and often induces increasing production costs (at least in the first years after conversion), this step includes high technical, market-related and financial risks for farmers. In order to reduce uncertainty and risks associated with such a conversion process, it does not seem to be sufficient solely to reduce transaction costs on the market by introducing standards for organic products and inspection systems for controlling these standards (as was the case in the USA), but additional financial incentives seem to be necessary to foster such a development – as was implemented in the EU. It can be questioned how long such direct payments should be given to farmers willing to convert to organic agriculture.

On the consumer markets, there is a rapidly growing demand for organic products in the USA and the EU. Although there are some difficulties concerning data availability, both markets are estimated to be around €10–13 billion, with significant regional differences (Marra, 2003; Yussefi and Willer, 2003). In the EU, 'mature' organic markets in countries such as Austria, Denmark, Sweden and Finland (with high per capita sales of organic products) can be observed parallel to 'emerging' organic markets (e.g. in Greece and Portugal) (Table 10.1). In most EU countries, organic plant products achieve higher market shares compared with organic livestock products. For future growth of the organic market, clear and reliable standards for organic products and corresponding consumer information, increasing interest and activities of the processing industry as well as increasing sales in conventional supermarkets are regarded as key success factors. In this sense, the US and EU regulations provided the starting point for innovation activities in the organic food area which were rewarded by consumers in recent years.

Empirical Results of a Survey of EU Food Processing Companies

In the following part of this chapter, empirical results from a survey of EU food manufacturing firms are presented which assess the current regulatory framework in the EU and its impact on innovation activities. In this context, we differentiate only between the food and the other industry sectors covered in the survey, in order to characterize the specifics of the food sector. In total, we received completed questionnaires from more than 260 companies, 20 of which were food companies. The companies were randomly selected from the commercial database of Dun & Bradstreet. At first, the companies were approached via fax or mail and asked either to fill in the questionnaire online or to download a pdf file and return the questionnaire via mail or fax. Secondly, around 2000 companies active in six selected industry sectors in further member states and 1000 companies in the USA and Canada were sent a paper version of the questionnaire. The response rate was rather low compared with other non-mandatory business surveys in Europe, because the complex issue of regulation and innovation within companies is not yet broadly taken into account. Consequently there is very often no one explicitly responsible for

Table 10.1.   Organic products as a percentage of the total sales value of selected products in EU member states in 2000.

	Cereals	Potatoes	Vegetables	Fruits	Milk	Beef	Eggs
Austria	10.2	4.8	6.1	3.8	7.7	3.5	2.7
Belgium	1.5	2.7	1.1	0.8	1.6	0.8	0.6
Denmark	14.5	8.3	12.8	3.2	13.1	3.0	11.9
Finland	5.5	2.3	7.1	1.6	0.6	1.0	2.3
France	1.7	1.9	–	–	0.9	0.4	2.2
Germany	4.7	3.5	3.6	1.8	1.3	3.1	1.9
Greece	0.2	0.0	0.3	–	–	–	0.0
Ireland	–	–	–	–	–	–	–
Italy	5.3	–	0.3	2.0	0.5	0.1	0.6
Luxembourg	3.0	1.3	0.4	0.3	–	0.6	–
Portugal	–	–	0.9	1.2	–	–	–
Spain	–	–	–	–	–	–	–
Sweden	9.7	4.8	4.3	0.9	1.8	1.0	2.1
The Netherlands	1.2	6.0	3.4	0.5	1.6	0.8	2.5
UK	1.6	0.7	4.4	1.9	1.1	0.4	2.5

Source: Hamm *et al.* (2002).

this issue. In more than one-third of the cases, the chief executive officer answered the questionnaire. The other answers stem from members of the R&D or the marketing department. Members of the legal department were hardly included in answering the questionnaire.

In Fig. 10.1, we present a differentiation of the importance of regulations as hampering factors for innovation. Although governmental regulations are more severe obstacles for innovation than non-governmental regulations, we find in the food sector the highest values and the largest discrepancies for non-governmental regulations. Furthermore, in general, the implementation of governmental regulations is a slightly more severe problem than the regulations themselves. Since there are rather different types of regulations, we differentiated more precisely between different types of regulations relevant for the development and introduction of new products and services. Figure 10.2 confirms that especially labelling and environmental regulations have a higher relevance for the food companies compared with the rest of the sample.

Regulation may be a hampering factor for the development and market introduction of new products and services. However, regulation may also have a positive influence on these activities, by forcing firms to search for new products and services. Furthermore, regulation may create new opportunities for new products and services by providing an adequate framework. Based on this background, we asked the companies to assess whether those regulations in the direct context of introducing new products and services into the market which have been assessed as important are more likely to have positive or negative impacts on these aspects. In general, Fig. 10.3 confirms that the regulations relevant for the introduction of new products in the food sector have more negative and less positive impacts than for the rest of the sample. They have both positive impacts on legal security including liability claims, and they have more detrimental impacts on domestic and foreign market shares and on the costs and risks of introducing new products into the market. These results raise serious concerns regarding the regulatory framework for the food sector in the

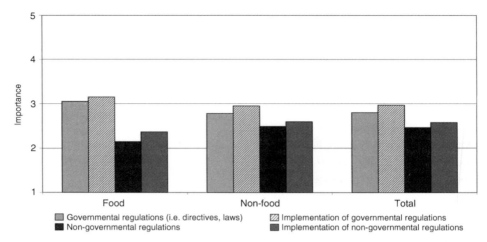

Fig. 10.1.   Barriers to innovation (1 = very low importance to 5 = very high importance).
Source: Company survey of Fraunhofer ISI, 2003.

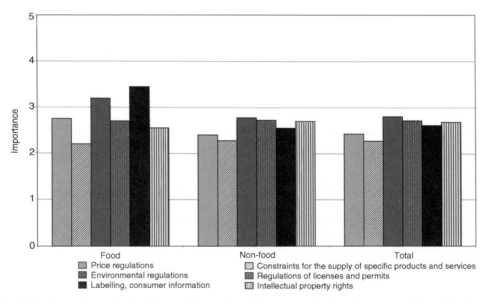

Fig. 10.2.   Importance of selected regulations relevant for new products and services differentiated by
sector (1 = very low importance to 5 = very high importance).
Source: Company survey of Fraunhofer ISI, 2003.

EU relevant for the introduction of new products. Without assuming a causality, it has to be mentioned that the innovativeness of the European food industry is below the innovative performance of the average of the European manufacturing sector.

In order to gain more precise information about the reasons for the negative impacts of the regulatory framework on innovations in the food sector, the level of agreement on some regulation-relevant key issues was collected by the respondents (see Fig. 10.4). In contrast to the rest of the sample, the companies in the food sector broadly share the experience that regulation hinders the development and market introduction of new products and

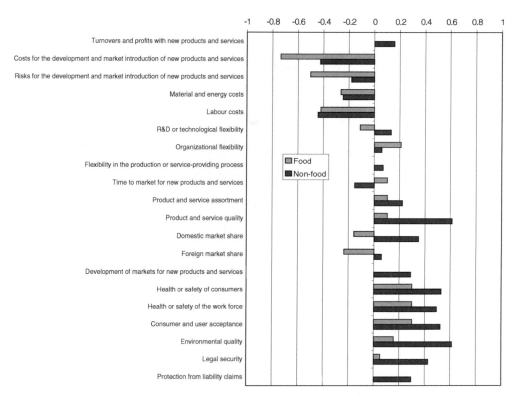

Fig. 10.3.   Impacts of the regulatory framework relevant for the introduction of new products and services (−1 = negative impact to +1 = positive impact).
Source: Company survey of Fraunhofer ISI, 2003.

services on the one hand, and disagree regarding the positive aspect of regulations creating opportunities for new products and services. Furthermore, the number of regulations is perceived as both too high and too non-transparent by almost all companies in the food sector. The majority of the food companies also have the perception that the regulations do not correspond to the needs of their customers. Regarding the implementation aspect of the regulatory framework in the food sector, it is evident that the approval procedures (time and cost) are not satisfactory, but the transparency and flexibility of the implementation also need to be improved.

Finally, the respondents were invited to give their opinion on proposals in order to improve the regulatory framework into a more innovation-friendly system. For the food companies a 'one-stop shop' solution regarding regulatory issues seems to be most adequate (see Fig. 10.5), and secondly the level of detail of governmental regulations should

be reduced, which is in line with using simpler language. In addition, the general amount of governmental regulations should be reduced. However, new regulatory bodies also obviously do not represent a solution for the food sector. This list of proposals should be regarded only as a starting point for more in-depth investigations, which are required to improve the efficiency of the specific regulatory frameworks relevant for introducing novel, functional or organic food products.

Role of Regulations for R&D and Marketing of New Products in the Food Industry

In recent years, the focus of innovation activities in the food industry has shifted from technical developments in their supply industries to

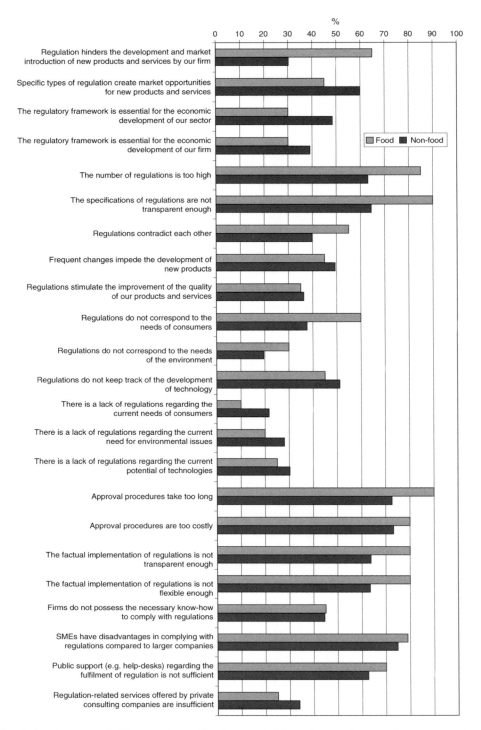

Fig. 10.4. Assessment of the current regulatory framework (share of companies agreeing as a percentage). Source: Company survey of Fraunhofer ISI, 2003.

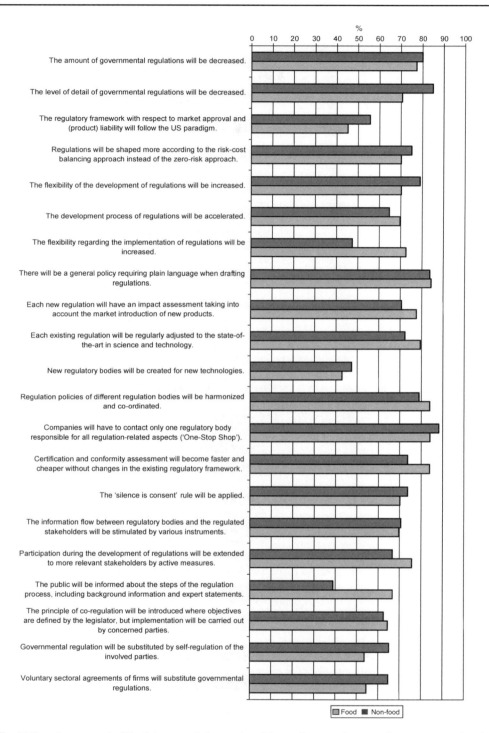

Fig. 10.5. Assessment of the future regulation system (share of companies agreeing as a percentage).
Source: Company survey of Fraunhofer ISI, 2003.

demand-oriented innovations, which results in a high number of new or modified products which are often combined with process innovations. In the coming years, the agro-food sector will be confronted with multiple new scientific approaches and technical opportunities which often have an interdisciplinary character. However, in particular SMEs of the food industry are not well prepared to profit from these developments. Therefore, the creation and building-up of interfacing competencies as well as the establishment of new external knowledge and competence networks seems to be of strategic relevance for many companies of the EU food industry. In this context it is advisable to widen the knowledge base of external cooperations and include clients, retail companies, research institutes, specialized service companies as well as other companies of the food and supply industries in such networks.

As illustrated in the case of functional food, in many innovative fields with relevance for the food industry the political and regulatory framework conditions in the EU often do not keep pace with scientific and technical discoveries or developments on the demand side. This relates in particular to regulatory aspects in which intensive discussions and coordination activities are required between the different member states. Such a situation of legal uncertainty or non-harmonized regulatory conditions between the different member states often impedes innovation activities and may result in loss of market opportunities. In this sense, a clear and harmonized regulatory framework in innovative fields seems to be a necessary but not sufficient prerequisite for the development of new products or services. Such a situation of legal and regulatory certainty is in the interests of both industrial companies (as a basis for commercial activities) and consumers (in particular for 'credence goods'). In this sense, there is a need for clarification, harmonization and implementation of regulations in the EU in particular for those innovative fields with relevance for the food industry in which consumers are interested in the respective products.

In particular, in highly interdisciplinary oriented innovation fields of the food industry (such as functional food), the institutional organization and administrative responsibilities impede innovation activities, since differing competent authorities with varying decision-making processes and procedures are responsible for the implementation, administration and control of existing regulations. In this sense, scientific and technical innovations require organizational changes which often take place with significant time delays. This is valid both for the EU and for the member states. Therefore, a more flexible framework for regulations should be created for newly emerging innovation fields which can be jointly formed by public authorities and early innovators.

As shown in the case study on organic food products, the definition of standards and the creation of labelling and control procedures does not seem to be sufficient for an early and fast take-off on the supply side. Although with these activities transaction costs are reduced on the market side, the high technical and market-related risks impede farmers from converting conventional farms to organic agriculture. In such a case, time-restricted and adopted final incentives seem to be an adequate instrument to speed up the adoption of an innovation. Such financial payments can be justified because positive environmental effects and reduced costs for financing conventional agricultural production are expected by increasing the rate of organic farming.

The analyses in the case studies have identified the creation of interfacing competencies as one of the most relevant tasks for SMEs of the EU food industry in order to carry out innovation activities successfully in future. Therefore, national and international policies should not solely concentrate on stimulating knowledge generation with relevance for the food industry, but should have the additional target of supporting advances of the knowledge base of the food industry companies themselves. In this sense, political activities should be targeted more strongly to the diffusion of new scientific approaches and technologies in the food industry rather than exclusively to the support of knowledge generation.

Acknowledgements

We would like to thank the European Commission (DG Enterprise/Innovation Policy Unit) for the financial support of the study as well as the cooperative interaction while performing the project in 2002 and 2003. The final report of this study has been published by the European Commission (Blind *et al.*, 2004).

References

Agriculture and Biotechnology Strategies (Canada) (2004) Global status of approved genetically modified plants. http://64.26.172.90/agbios/dbase.php?action=Synopsis

Blind, K., Bührlen, B., Kotz, C., Menrad, K. and Walz, R. (2004) *New Products and Services: Regulation Shaping New Markets*. Report to DG Enterprise of the European Commission. http://www.cordis.lu/innovation-policy/studies/gen_study11.htm

Centrale Marketing-Gesellschaft der Deutschen Agrarwirtschaft mbH (2002) *Functional Food – ein Regionalvergleich*. CMA, Bonn, Germany.

Codex Alimentarius (2003) Current official standards. http://www.codexalimentarius.net/standard_list.asp

Dustmann, H. and Weindlmaier, H. (2002) *Delphi-Studie zur Untersuchung der Herstellungs- und Absatzbedingungen funktioneller Lebensmittel in Deutschland*. Forschungszentrum für Milch und Lebensmittel Weihenstephan, Technische Universität München, München, Germany.

Esposti, R. and Sorrentino, A. (2002) *Policy and Regulatory Options on Genetically Modified Crops: Why USA and EU Have Different Approaches and How WTO Negotiations Can be Involved?* Paper presented at the 6th ICABR Conference, July 11–14, 2002, Ravello, Italy.

Eurobarometer (2001) *GMOs and BSE – The European's View*. Press release of December 2001. http://www.europa.eu.int/comm/research/press/2001/pr2312en.html

European Commission (1997) *Green Paper. The General Principles of Food Law in the European Union*. http://europa.eu.int/scadplus/leg/en/lvb/l21220.htm

European Commission (2000) White paper on food safety. http://www.europa.eu.int/scadplus/leg/en/lvb/l32041.htm

European Commission (2003a) Commission recommendation of 23 July 2003 on guidelines for the development of national strategies and best practices to ensure the co-existence of genetically modified crops with conventional and organic farming. European Commission, Brussels.

European Commission (2003b) Proposal for a regulation of the European Parliament and of the Council on nutrition and health claims made on foods. http://europa.eu.int/eur-lex/en/com/pdf/2003/com2003_0424en01.pdf European Commission, Brussels.

Food and Drug Administration, Department of Health and Human Services (1992) Statement of policy: foods derived from new plant varieties. *Federal Register* 57, 22984–23001.

Food and Drug Administration, Department of Health and Human Services (2001) Pre-market notice concerning bioengineered foods. *Federal Register* 66, 4706–4738.

Garza, C. and Stover, P. (2003) General introduction: the role of science in identifying common ground in the debate on genetic modification of foods. *Trends in Food Science and Technology* 14, 182–190.

Gaskell, G., Allum, N., Bauer, M., Durant, J., Allansdottir, A., Bonfadelli, H., Boy, D., de Cheveigné, S., Fjaestad, B., Gatteling, J.M., Hampel, J., Jelsoe, E., Correia Jesaino, J., Kohring, M., Kronberger, N., Midden, C., Nielsen, T.H., Przestalski, A., Rusanen, T., Sakellaris, G., Torgersen, H., Twardowski, T. and Wagner, W. (2000) Biotechnology and the European public. *Nature Biotechnology* 18, 935–938.

Greenberg, E.F. (2000) Public policy issues. In: Brody, A.L. and Cord, J.B. (eds) *Developing Food Products for a Changing Market Place*. Technomic Publishing Company, Lancaster, UK, pp. 465–475.

Greene, C. and Kremen, A. (2003) *US Organic Farming in 2000–2001: Adoption of Certified Systems*. Agriculture Information Bulletin No. 780. USDA Economic Research Service, Resource Economics Division, Washington, DC.

Groeneveld, M. (2000) *Funktionelle Lebensmittel – 2. Dokumentation zur aktuellen wissenchaftlichen Diskussion*. Institut für Lebensmittelwissenschaft und -information GmbH, Bonn, Germany.

Hamm, U., Gronefeld, F. and Halpin, D. (2002) *Analysis of the European Market for Organic Food*. University of Wales Aberystwyth, School of Management and Business, Aberystwyth, UK.

Heasman, M. and Mellentin, J. (2001) *The Functional Foods Revolution. Healthy People, Healthy Profits?* Earthscan Publications, London.

Hilliam, M. (2000) Functional food – how big is the market? *World of Food Ingredients* 12, 50–52.

James, C. (2003) *Global Status of Commercialized Transgenic Crops 2003*. International Service for the Acquisition of Agri-biotech Applications (ISAAA).

Lampkin, N., Foster, C., Padel, S. and Midmore, P. (1999) The policy and regulatory environment for organic farming in Europe. In: Dabbert, S., Lampkin, N., Michelsen, J. Nieberg, H. and Zanoli, R. (eds) *Organic Farming in Europe: Economics and Policy*. Vol. 1. Department of

Farm Economics, University of Hohenheim, Stuttgart, Germany.

Lheureux, K., Libeau-Dulos, M., Nilsagard, H., Rodriguez Cerezo, E., Menrad, K., Menrad, M. and Vorgrimler, D. (2003) Review of GMOs Under Research and Development and in the Pipeline in Europe. Institute for Prospective Technological Studies, Seville/Fraunhofer Institute for Systems and Innovation Research, Karlsruhe.

Loureiro, M.L. (2003) GMO food labelling in the EU: tracing 'the seeds of dispute'. *EuroChoices* 2, 18–23.

Marra, J. (2003) Consumer health and wellness attitudes. *Innova* 5/6, 16–18.

Menrad, K. (2001) Innovations at the borderline of food, nutrition and health in Germany – a systems' theory approach. *Agrarwirtschaft* 50, 331–341.

Menrad, K. (2003a) Market and marketing of functional food in Europe. *Journal of Food Engineering* 56, 181–188.

Menrad, K. (2003b) New products and services. Analysis of regulations shaping new markets. The impact of regulations on the development of new products in the food industry. Fraunhofer Institute for Systems and Innovation Research (ISI), Karlsruhe http://www.isi.fhg.de/ti/Downloads/KB_Case_Study_Food.pdf

Menrad, M., Hüsing, B., Menrad, K., Reiß, T., Beer-Borst, S. and Zenger, C.A. (2000) *Functional Food. No. TA 37/2000*. Schweizerischer Wissenschafts- und Technologierat, Bern, Switzerland.

Organic Centre Wales (2003b) Certified and policy-supported organic and in-conversion land area in Europe (ha). http://www.organic.aber.ac.uk/statistics/euroarea.htm

Otsuka, Y. (2003) Socioeconomic considerations relevant to the sustainable development, use and control of genetically modified foods. *Trends in Food Science and Technology* 14, 294–318.

Reiss, T. and Dominguez Lacasa, I. (2003) Performance of European member states in biotechnology. EPOHITE workshop dated June 17, 2003 in Paris. http://www.epohite.fhg.de/Documents/Wp4_final_workshop/performance.pdf

Ringel Heller, I. and Silverglade, B. (2001) Functional foods – health boom or quakery? *Bundesgesundheitsblatt-Gesundheitsforschung-Gesundheitsschutz* 3, 214–218.

Sauter, A. and Meyer, R. (2000) *Risikoabschätzung und Nachzulassung-Monitoring transgener Pflanzen*. TAB-Arbeitsbericht No. 68. Büro für Technikfolgen-Abschätzung beim Deutschen Bundestag, Berlin.

Storlie, J. and Brody, A.L. (2000) Mandatory food package labeling in the United States. In: Brody, A.L. and Cord, J.B. (eds) *Developing Food Products for a Changing Market Place*. Technomic Publishing Company, Lancaster, UK, pp. 409–438.

United States Department of Agriculture (2003) *Farms and Land in Farms*. USDA, National Agricultural Statistics Service, Washington, DC.

Yussefi, M. and Willer, H. (eds) (2003) The *World of Organic Agriculture: Statistics and Future Prospects 2003*. International Federation of Organic Agriculture Movements (IFOAM), Tholey-Theley, Germany.

11 Patents versus Plant Varietal Protection

Derek Eaton and Frank van Tongeren

Agricultural Economics Research Institute (LEI), Wageningen University and Research Centre (WUR), PO Box 29703, 2502 LS The Hague, The Netherlands

Introduction

This chapter develops a theoretical model to analyse the impacts of intellectual property rights (IPRs) policy options on innovation in the plant breeding sector and social welfare. One motivation is the emerging world of differentiated IPR policies in agricultural biotechnology. The USA and the European Commission (EC) are pursuing separate policies with respect to the patenting of innovation in the plant breeding sector. In contrast to the stronger patenting approach of the USA, the EC Directive 98/44/EC on the legal protection of biotechnological inventions only allows plant breeders' rights (PBRs) for plant varieties, while patent protection is to be available for biotechnological inventions such as the use of genetic transformation techniques in plants. Many developing countries and economies in transition, in the fulfillment of their TRIPS obligations, have passed PBR legislation, only some of which is UPOV-compliant (see below), and are now in the stages of institutional implementation. These countries are generally less advanced in their corresponding patent obligations for biotechnology inventions.

A second motivation is the proposal by some large plant breeding companies in the private sector to increase the scope of PBR protection in Europe, the USA (despite its relatively lower importance there as a form of protection) and elsewhere, through adjustments to the UPOV Convention.

The International Union for the Protection of New Varieties of Plants (UPOV) is an international agreement on the technical requirements for PBR protection as well as the resulting scope of protection. Both of the two most recent versions of the treaty, referred to as the 1978 Act and the 1991 Act, include a 'breeders' exemption' which means that protected varieties may be used by competitors as material in their breeding programmes, without any obligation to the original right holder.[1] Recently, Pioneer Hi-Bred proposed a phasing-in of the breeders' exemption after a certain number of years, determined per crop according to factors such as the length of the breeding cycle and the product lifetime.[2] This period would obviously be shorter than the duration of PBR protection, e.g. 10 years. Under such an arrangement, competing breeders would have to wait for these first 10 years to pass before being able to use a protected variety in their breeding programmes, without permission of the right holder.

The alternatives for IPR protection of plant varieties can now be viewed as a continuum from PBRs to patents. In between these two options, a phased-in breeders' exemption would be an intermediate option to strengthening the scope of protection of PBRs. At the extreme, if the breeders' exemption is not phased in until the expiry of the PBR, then the protection is quite comparable with that of patents, ignoring the requirements for obtaining protection.

The question for policymakers is what is the effect of increased protection on innovation incentives, including the indirect effects arising from changes in market structure. Ultimately, policy should be chosen to maximize the resulting welfare outcome. However, another policy objective has also been added to the list recently and concerns the use and maintenance of genetic diversity. On the one hand, increasing concentration that could result from broader IPR protection could lead to a more limited range of crop varieties being available. On the other hand, for crops such as maize, Pioneer argues that increased appropriation by breeders of benefits is necessary to finance the greater R&D investments in order to incorporate a broader range of genetic material, such as wild relatives, into their breeding programmes (Donnenwirth et al., 2004).

In this chapter, we develop a simple theoretical model to attempt to illuminate some of the trade-offs between innovation incentives and product quality/diversity, arising from the effects of changes in IPR protection. Our model is a simple adaptation of a standard model elaborated by Motta (1993) of vertical product differentiation with endogenous quality choice. The following section briefly reviews some relevant strands of literature on the scope of IPR protection. We then present the theoretical model in which we interpret a phased-in breeders' exemption as affecting the cost functions for breeders. Numerical simulations in the subsequent section illustrate how a phased-in breeders' exemption may drive competitors out of the market while still leaving the leading firm with little incentive to increase investment. In the concluding remarks, we comment on the possibilities of undertaking further empirical and theoretical research on this issue.

Approaches to Analysing the Scope of IPR Protection

In this section, we review relevant literature on the scope of IPR protection. There have been relatively few attempts to model PBRs as an explicit form of IPR (Lesser, 1997). Alston and Venner (2002) developed a model of partial appropriability for a monopolistic breeding sector with the breeders' exemption mentioned as one of the explanations for the relatively weak appropriability provided by

PBRs in the USA. Patents, in contrast, have received considerable attention from economists in terms of both theoretical and empirical research. The issue of interest here is to what extent the findings of this research are applicable for the phased-in breeders' exemption issue.

The option of phased-in breeders' exemption creates a continuum of increasing protection between PBRs and patents. This concept is similar to the breadth of patent protection. Breadth is one dimension of the scope of patent protection. O'Donoghue (1998) distinguishes between lagging breadth, which describes the extent of protection against imitators, and leading breadth, which is the extent of protection against subsequent innovators.[3] It is possible to interpret a phased-in breeders' exemption as extending the breadth of patent-like protection. Broader protection means that subsequent breeders must invest more to develop a variety with greater benefits for farmers, without infringing the existing variety.

The last 15 years have seen a steady progression in the complexity of models developed to analyse the breadth of patent protection. Earlier work concentrated on a two-stage framework and the need for broad patent protection and licensing provisions to transfer benefits from a subsequent innovator back to a first innovator (e.g. Scotchmer, 1991, 1996). More recent work has extended the analysis to an infinite-stage setting of sequential innovation, highlighting a need to balance upstream and downstream incentives (O'Donoghue et al., 1998; Bessen and Maskin, 2000) and making a link with the quality ladders framework of endogenous growth (O'Donoghue, 1998).

The patent scope literature has not yet examined all the important aspects of the problem. Market structure and power are not captured in such frameworks where the most recently developed product enjoys a monopoly position until replaced by the subsequent generation. A good deal of the debate surrounding the PBR–patent issue in plant breeding concerns the potential effects of broader IPR protection in terms of greater concentration and the ability of other firms to continue competing with a leader. More specifically, a phased-in breeders' exemption, or even patent protection, could result in a carving up of the germplasm pool among a very limited number of breeders remaining in the market.

This germplasm pool issue reflects a specific characteristic of the plant breeding sector.[4]

Plant breeding differs from many other forms of cumulative innovation in that further innovation (breeding) is not possible without physical access to previous innovations. This is arguably the principal reason why PBRs were developed in the first place, as an alternative to utility patents for which publication releases (in principle) the knowledge behind the innovation to competitors and researchers. There is no parallel to this information disclosure with PBR as it is effectively included only in the variety's genetic sequence.

There are various approaches available for representing the germplasm pool issue. Spillovers in R&D offer one possibility that could be further explored, building on the d'Aspremont and Jacquemin (1988) framework (see also Amir, 2000, for a summary of recent developments). There is also a potential link to make with the use of spillovers in Kortum's (1997) quality ladder model of endogenous growth. Monopolistic competition may offer another approach.

In this chapter, we choose a vertical product differentiation framework in order also to incorporate a representation of varietal diversity, as well as the notion of absolute improvements to plant varieties. A range of available plant varieties is beneficial to the extent that farms are heterogeneous. Furthermore, a greater number of plant varieties is likely to entail a broader range of genetic material being used, which helps attain conservation objectives.[5]

A well-known approach to modelling a duopoly in which firms endogenously choose their quality levels was developed by Shaked and Sutton (1982). With symmetric firms, they show that the equilibrium solution will entail one firm supplying a higher quality than the other. Their analysis was restricted to where costs of quality improvements were fixed. Motta (1993) allows variable cost functions for quality and analyses Bertrand versus Cournot competition, confirming that product differentiation arises in a symmetric duopoly. In the model presented below, we modestly extend Motta's model.

An Oligopolistic Model of Plant Breeding

To analyse the effects of restricting the breeders' exemption, we develop a model of vertical product differentiation for the plant breeding sector in which two plant breeding firms choose the quality

and then the price of their respective seed varieties in a two-stage game. This model of endogenous quality choice is an extension of the study by Motta (1993). The difference here is that we allow for asymmetric cost functions at the R&D stage as a means of representing the effects of a phased-in breeders' exemption. As will be seen below, our logic for such an approach is based on the representation of plant breeding within a search-theoretic framework by Evenson (1998).

We begin with the basic structure of the model. For simplicity, the farming sector is modelled analogously to consumers in a model of vertical product differentiation. Farms compete in a competitive output market as price takers. Seed is their only input and their profit is given by $V = vu - p$, where u and p are, respectively, the quality and price of seed. Farms differ in the characteristics of their land and local growing conditions which are captured in the parameter $v \in [v_{min}, v_{max}]$, with v being uniformly distributed with unit density. As is typical with such models, we normalize the quantity purchased such that each farm buys one unit of seed unless $vu - p < 0$, in which case they neither purchase nor produce.[6]

We assume that there are only two firms that play a two-stage game, with each firm producing one variety of seed. In the first stage, firms must choose the quality, u_1 and u_2, respectively, of their variety, with $u_1, u_2 \geq 1$. This lower bound of quality could be interpreted as a legal minimum standard (e.g. seed certification requirements) or as the existing quality level on which firms have to improve. We interpret this first stage as an R&D stage in which firms incur fixed costs to achieve a chosen quality level. In the second stage, Bertrand price competition takes place. For simplicity and without loss of generality, we assume that the costs of producing seed are zero. A subgame perfect Nash equilibrium is solved through backward induction.

Firm 1 is the quality leader and firm 2 is a follower: $u_1 \geq u_2$. Farm v_{12} earns equal profits from variety 1 as it does from variety 2, so $v_{12} = (p_1 - p_2)/(u_1 - u_2)$. Farm v_{02} earns zero profits from variety 2: $v_{02} = p_2/u_2$. Thus farms distributed between $v_{12} \leq v \leq v_{max}$ will purchase variety 1; and farms distributed between $v_{02} \leq v \leq v_{12}$, variety 2. This leads to the following simple inverse demand functions:

$$q_1 = v_{max} - (p_1 - p_2)/(u_1 - u_2)$$

$$q_2 = (p_1 - p_2)/(u_1 - u_2) - (p_2/u_2)$$

(1)

Given their quality choices (u_1, u_2), firms seek to maximize their profits, $\Pi_i = p_i q_i$, in the second stage by setting price. The first order conditions for profit maximization are:

$$\partial \Pi_1 / \partial p_1 = v_{max} + (p_2 - 2p_1)/(u_1 - u_2) = 0$$
$$\partial \Pi_2 / \partial p_2 = (p_2 - 2p_1)/(u_1 - u_2) - 2p_2/u_2 = 0 \qquad (2)$$

Solving for equilibrium prices yields,

$$p_1 = 2v_{max} u_1 (u_1 - u_2)/(4u_1 - u_2)$$
$$p_2 = v_{max} u_2 (u_1 - u_2)/(4u_1 - u_2) \qquad (3)$$

and profits in the second stage are:

$$\Pi_1(u_1, u_2) = 4(u_1 - u_2)[v_{max} u_1 (4u_1 - u_2)]^2$$
$$\Pi_2(u_1, u_2) = u_1 u_2 (u_1 - u_2)[v_{max}/(4u_1 - u_2)]^2 \qquad (4)$$

The first stage quality game is then solved by firms choosing their quality levels to maximize profits which now incorporate the cost functions. This R&D cost function is taken to be a quadratic function of quality. This was the choice of Motta and is also a common functional form in the literature on R&D. We feel that in this case, its choice can also be justified by the work of Evenson and Kislev (1976) and Evenson (1998) on plant breeding production functions. Placing plant breeding within a search-theoretic framework, Evenson proposes that the expected value, or productivity improvement, z, obtained from a draw, or search, among a population of n varieties, or genebank accessions, can be approximated[7] as $E(z) = a + b \cdot \ln(n)$. Costs can be viewed as being proportional to n, the size of the search population. This production function thus corresponds to a cost function such as the exponential function, in which marginal costs are increasing at an increasing rate in the quality or trait being sought. For simplicity, we use a quadratic cost function, which has constantly increasing marginal costs. For our purposes, this is a conservative approach as a steeper function can be expected to reinforce the strength of the results.

Whereas Motta (1993) used identical cost functions for both firms, we now introduce an extra parameter to account for effects of restricting or phasing in the breeders' exemption. This leads to asymmetric costs, with firm 2 experiencing higher costs as a result of restricted or delayed access to the new variety of firm 1. Firm 1's costs

are given by $\lambda u_1^2/2$ and firm 2's costs by $\alpha \lambda u_2^2/2$. In addition to $\alpha(>1)$ for the breeders' exemption, λ is a variable that allows us to examine the effects of proportionally increasing the cost functions of both firms. The motivation for this is the exhaustion of the genetic pool issue mentioned above; as breeders exhaust the existing genetic pool in which they are searching, either recharge will become necessary, through costly germplasm acquisition and evaluation, or genetic material may be sought in other species, including wild relatives.

In the first stage of the game, firms now choose their quality levels, u_1 and u_2, respectively, to maximize their full profit functions:

$$\pi_1(u_1, u_2) = 4(u_1 - u_2)[v_{max} u_1 (4u_1 - u_2)]^2$$
$$\qquad\qquad - \lambda u_1^2/2$$
$$\pi_2(u_1, u_2) = u_1 u_2 (u_1 - u_2)[v_{max}/(4u_1 - u_2)]^2$$
$$\qquad\qquad - \alpha \lambda u_2^2/2 \qquad (5)$$

The first order conditions are,

$$\partial \pi_1 / \partial u_1 = 4u_1 v_{max}^2 [4u_1^2 - 3u_1 u_2 + 2u_2^2/(4u_1 - u_2)]^3$$
$$\qquad\qquad - \lambda u_1 = 0$$
$$\partial \pi_2 / \partial u_2 = v_{max}^2 u_1^2 (4u_1 - 7u_2)/(4u_1 - u_2)^3$$
$$\qquad\qquad - \alpha \lambda u_2 = 0 \qquad (6)$$

Substituting for v_{max} and rearranging,

$$4u_1^3 - (7 + 16\alpha)u_1^2 u_2 + 12\alpha u_1 u_2^2 - 8\alpha u_2^3 = 0 \qquad (7)$$

As $u_1 \geq u_2$, let $u_1 = \mu \cdot u_2$, with $\mu \geq 1$. Substituting,

$$4\mu^3 - (7 + 16\alpha)\mu^2 + 12\alpha\mu - 8\alpha = 0 \qquad (8)$$

For given values of α, Equation 8, a third-order polynomial in μ, can then be solved using numerical methods. Given μ, we can find all the parameters in the model, expressed as a function of v_{max}:

$$u_2 = \frac{\mu^2(4\mu - 7)}{\alpha\lambda(4\mu - 1)} v_{max}^2 \qquad u_1 = \mu u_2 \qquad (9)$$

$$v_{12} = \frac{(2\mu - 1)}{(4\mu - 1)} v_{max} \qquad v_{02} = \frac{(\mu - 1)}{(4\mu - 1)} v_{max} \qquad (10)$$

Equilibrium prices are given in Equation 3, and quantities by,

$$q_1 = 2\mu v_{max}/(4\mu - 1) \qquad q_2 = \mu v_{max}/(4\mu - 1) \qquad (11)$$

Profits of firm 1 and 2 are given above in Equation 5. Farm profits are measured with a simple

Table 11.1. Benchmark results ($\alpha = 1$ and $\lambda = 1$).

	Firm 1	Firm 2	Farms
u	$0.2533 \cdot v_{max}^2$	$0.0482 \cdot v_{max}^2$	—
p	$0.1077 \cdot v_{max}^3$	$0.0103 \cdot v_{max}^3$	—
q	$0.5250 \cdot v_{max}$	$0.2625 \cdot v_{max}$	$0.7875 \cdot v_{max}$
π	$0.0244 \cdot v_{max}^4$	$0.0015 \cdot v_{max}^4$	$0.0432 \cdot v_{max}^4$
Total profits	$0.0692 \cdot v_{max}^4$		
v_{02}	$0.2125 \cdot v_{max}$		
v_{12}	$0.4750 \cdot v_{max}$		

Source: authors' own calculations.

formula, as is typically done with consumer surplus (see Motta, 1993):

$$\pi_{Farms} = \int_{v_{02}}^{v_{12}} (vu_2 - p_2)\,dv + \int_{v_{12}}^{v_{max}} (vu_1 - p_1)\,dv \qquad (12)$$

The proof that Equation 8 represents a Nash equilibrium follows the proof of Motta (1993) and depends on the numerical solutions for μ. Our use of the additional parameters, α and λ, only reinforces the logic of the proof, which is not presented here to conserve space but is available upon request. The following section examines how the solution varies for different values of α and λ.

Simulations

Numerical simulations are conducted for various values of α and λ. We begin with the benchmark solution where $\alpha = 1$ and $\lambda = 1$, i.e. with no asymmetry in costs, as also calculated by Motta (1993). These are summarized in Table 11.1 as coefficients on v_{max}. The equilibrium quality levels are $u_1 = 0.2533 \cdot v_{max}^2$ and $u_2 = 0.0482 \cdot v_{max}^2$ which are related by the factor $\mu = 5.2512$.

Given the stylistic nature of the model, equilibrium parameter values were calculated for a range of $\alpha = \{1.0, 1.1, \ldots, 10.0\}$, with higher α increasing proportionally the cost function of firm 2. This leads to a seemingly linear increase in μ, as seen in Fig. 11.1 from 5.2512 for $\alpha = 1$ to 41.031 for $\alpha = 10$.[8]

What happens to the qualities offered by both firms as it becomes more difficult for firm 2 to conduct its R&D? Figure 11.2 shows that firm 2

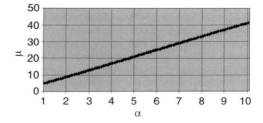

Fig. 11.1. μ for increasing α.

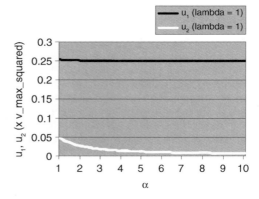

Fig. 11.2. Equilibrium qualities for increasing α.

decreases its quality as a result of the higher costs. Recall that we interpret these higher breeding costs for firm 2 as resulting from a phased-in, or even eliminated, breeders' exemption. A doubling of firm 2's costs relative to that of firm 1 results in a quality decline for the former of roughly a half. As α increases to 5, firm 2's quality approaches zero (recall that there is a minimum quality required by the model). How does firm 1 react to the reduced

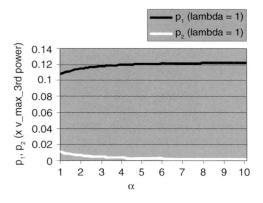

Fig. 11.3. Equilibrium prices for increasing α.

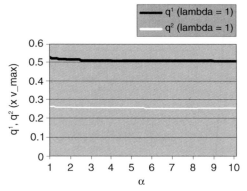

Fig. 11.4. Equilibrium quantities for increasing α.

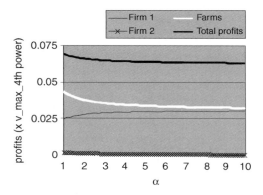

Fig. 11.5. Firm and farm profits.

threat posed by firm 2? Although perhaps difficult to see in Fig. 11.2, firm 1's quality also declines but only marginally, remaining roughly constant as α increases. Firm 1 does not therefore have any incentive in this setting to increase its R&D investment as a result of increased scope of IPRs and to develop a higher quality seed variety. Firm 1 can increase its price while moderately decreasing quality, thus tending towards monopolistic behaviour (Figs 11.3 and 11.4).

The results support the argument that increased scope of IPR protection will lead to market power for the leading seed breeder, at the expense of both competitors and client farms. This can be seen by examining the effects of α on profits as shown in Fig. 11.5. Firm 1's profit increases as α increases. Firm 2's profits are of a much lower magnitude and decrease even further. Thus, the combined profits of the two seed breeders increase while those of the farming sector decrease. The decline in

farm profits is greater than the increase in those of firm 1, and total profits, a measure of economic surplus in this partial framework, decrease. Thus a phased-in breeders' exemption would be welfare decreasing. The decrease in farm profits can be divided between two effects. First, firm 1, which is offering roughly the same quality, is able to increase its price. Secondly, farms in the lower end of the range of v suffer from the reduced quality of firm 2 which approaches the minimum possible quality level. In other words, many farms have effectively less choice for economically profitable seed varieties. In the modelling framework, this is seen by v_{02} (the farm, earning zero profits which is indifferent between variety 2 and not producing at all) increasing with α (not shown).

The equilibrium solutions of the model were also found for various values of λ between 1 and 10. Recall that λ allows for equal increases in costs of quality development for both firms, representing a general narrowing of the germplasm pool and an increase in the costs of achieving a given quality increase. With $\lambda > 1$, qualities decrease for both firms as do profits, but the effect of increasing α follows a similar pattern with results that are relatively comparable with those discussed above. This follows from the fact that λ factored out of Equation 8 above, meaning that values of μ do not change with $\lambda > 1$. There may, however, be better ways to capture increasing costs for the breeding sector as a whole in such a model. Quality is represented herein as a one-off choice, while the increasing costs refer to those necessary to achieve the subsequent, incremental quality improvement. In the next and final section, we discuss other limitations to the model and possible future directions for research.

Conclusions

In this chapter we have developed a model that illustrates the potential negative effects of increasing the scope of PBR protection on plant varieties, as proposed with either a phased-in breeders' exemption or patents. Such a policy could increase the costs for varietal development for breeding companies, in particular if their access to varieties of the market leader is constrained. We have represented this scenario as an asymmetrical increase in the costs of varietal development. This feature is incorporated into a simple extension to Motta's (1993) and Shaked and Sutton's (1982) model of endogenous quality choice under price competition, a simple model of a duopoly. This stylized model shows how the leading breeder is able to exercise market power while not increasing its own R&D investments or seed quality. The profits of the market leader increase, but this is more than offset by decreases to farm profits, which also lose from a decrease in quality levels offered. In addition to being detrimental for innovation incentives and welfare, we also argue that a phased-in breeders' exemption might also reduce the use of germplasm resources with negative diversity consequences.

Numerous issues could be studied further. Our stylistic model suffers from the usual deficiencies but opens up some interesting possibilities for more detailed analysis. In particular, it seems important to develop further approaches for examining how restricting access among firms to the genetic pool will affect their ability to breed new varieties, particularly for firms on the competitive fringe or operating as followers. Our analysis suggests the possibility of examining the implications for their breeding costs as an important line of empirical research. While the issues at stake are quite important, and include the use and maintenance of genetic diversity, empirical research would be difficult given the secrecy with which breeders guard information concerning the germplasm they are using. Pioneer has argued that this carving up of the genetic pool would be mitigated by cross-licensing of germplasm (Donnenwirth *et al.*, 2004). It could therefore

also be useful to apply lessons and modelling approaches from the extensive literature on licensing of patent-protected innovations.

Other extensions to the analysis include addressing multiproduct firms in oligopoly, as well as other market structures such as more than two firms, or some other forms of monopolistic competition. Another issue is whether there is some added value to a more direct representation of the breeding production function developed by Evenson (1998). Thirdly, different distributions of farm heterogeneity may be more realistic for some circumstances (e.g. developing versus industrialized country agriculture) and yield different results. Finally, a more explicit treatment of the consequences of the application of modern biotechnology to breeding, in terms of the production function (Evenson 1998) or the costs involved, would be interesting. At issue here is whether the scale economies involved provide the basis for natural oligopoly, or even monopoly, in the plant breeding sector.

Notes

[1] The 1991 Act of UPOV introduced the exclusion to the breeders' exemption for 'essentially-derived varieties' (EDV). Debate continues among UPOV members and also plant breeding companies as to an appropriate and practical definition of EDV. The basic idea is to ensure that a breeder subsequently combines a protected variety of a competitor together with other separate varieties, and does not simply introduce minor changes, which has become more feasible with genetic transformation techniques.

[2] Presentation by R. McConnell, CEO, Pioneer Hi-Bred at the International Seed Federation (ISF) Seminar on Intellectual Property Rights and Access to Plant Genetic Resources, held in Berlin, 27–28 May 2004; see also Donnenwirth *et al.* (2004).

[3] Denicolò and Zanchettin (2002) use the term, 'forward protection' for leading breadth. Earlier terms include 'height' (Klemperer, 1990; van Dijk, 1996). O'Donoghue (1998) provides a clear discussion of the relationship between these various concepts of patent breadth.

[4] The germplasm pool issue is also a manifestation of the 'carving up the commons' phenomenon (see Falcon and Fowler, 2002, for a detailed discussion with respect to plant breeding in an international context).

5 There is some controversy as to the extent to which *in situ* as well as *ex situ* conservation is necessary, as well as the extent to which diversity should be present in commercial fields, particularly if this involves a trade-off with (current) productivity (e.g. Wright, 1997). However, there is less controversy over the fact that conservation without use has little purpose.

6 We assume that the market is not covered, as in Motta (1993) who discusses the implications of assuming full or partial market coverage. Full market coverage would affect our specific numerical results but not their qualitative interpretation. Given the extent of competition in the seed sector and the predominance of farm-saved seed in many countries, partial market coverage seems somewhat more compelling. A possible extension is offered by Wauthy (1996) who generalizes the original model of Shaked and Sutton (1982) with endogenous determination of full or partial market coverage.

7 Evenson and Kislev (1976) showed that this was the case if the underlying process by which z is obtained is exponentially distributed. Kortum (1997) has shown that this also holds for most commonly used distributions except those that are 'fat-tailed'.

8 Given a seemingly linear relationship between α and μ, it may be possible to devise an analytical solution that eases further analysis.

References

Alston, J.M. and Venner, R.J. (2002) The effects of the US Plant Variety Protection Act on wheat genetic improvement. *Research Policy* 31, 527–542.

Amir, R. (2000) Modelling imperfectly appropriable R&D via spillovers. *International Journal of Industrial Organization* 18, 1013–1032.

Bessen, J. and Maskin, E. (2000) *Sequential Innovation, Patents, and Imitation.* 00/01. Department of Economics, Massachusetts Institute of Technology, Massachusetts.

d'Aspremont, C. and Jacquemin, A. (1988) Cooperative and noncooperative R&D in duopoly with spillovers. *American Economic Review* 78, 1133–1137.

Denicolò, V. and Zanchettin, P. (2002) How should forward patent protection be provided? *International Journal of Industrial Organization* 20, 801–827.

Donnenwirth, J., Grace, J. and Smith, S. (2004) Intellectual property rights, patents, plant variety protection and contracts: a perspective from the private sector. *IP Strategy Today* 9, 19–34.

Evenson, R.E. (1998) Plant breeding: a case of induced innovation. In: Evenson, R.E., Gollin, D. and Santaniello, V. (eds) *Agricultural Values of Plant Genetic Resources.* CAB International, Wallingford, UK, pp. 29–42.

Evenson, R.E. and Kislev, Y. (1976) A stochastic model of applied research. *Journal of Political Economy* 84, 265–282.

Falcon, W.P. and Fowler, C. (2002) Carving up the commons – emergence of a new international regime for germplasm development and transfer. *Food Policy* 27, 197–222.

Klemperer, P. (1990) How broad should the scope of patent protection be? *RAND Journal of Economics* 21, 113–130.

Kortum, S.S. (1997) Research, patenting, and technological change. *Econometrica* 65, 1389–1419.

Lesser, W. (1997) Assessing the implications of intellectual property rights on plant and animal agriculture. *American Journal of Agricultural Economics* 79, 1584–1591.

Motta, M. (1993) Endogenous quality choice: price vs. quantity competition. *Journal of Industrial Economics* 41, 113–131.

O'Donoghue, T. (1998) A patentability requirement for sequential innovation. *RAND Journal of Economics* 29, 654–679.

O'Donoghue, T., Scotchmer, S. and Thisse, J.F. (1998) Patent breadth, patent life, and the pace of technological progress. *Journal of Economics and Management Strategy* 7, 1–32.

Scotchmer, S. (1991) Standing on the shoulders of giants: cumulative research and the patent law. *Journal of Economic Perspectives* 5, 29–41.

Scotchmer, S. (1996) Protecting early innovators: should second-generation products be patentable? *RAND Journal of Economics* 27, 322–331.

Shaked, A. and Sutton, J. (1982) Relaxing price competition through product differentiation. *Review of Economic Studies* 49, 3–14.

van Dijk, T. (1996) Patent height and competition in product improvements. *Journal of Industrial Economics* 44, 151–167.

Wauthy, X. (1996) Quality choice in models of vertical differentiation. *Journal of Industrial Economics* 44, 345–353.

Wright, B.D. (1997) Crop genetic resource policy: the role of *ex situ* genebanks. *Australian Journal of Agricultural and Resource Economics* 41, 81–115.

12 Governing Innovative Science: Challenges Facing the Commercialization of Plant-made Pharmaceuticals

Stuart Smyth[1], George G. Khachatourians[2] and Peter W.B. Phillips[1]

[1]Department of Agricultural Economics and [2]Department of Applied Microbiology and Food Science, University of Saskatchewan, 51 Campus Drive, Saskatoon, Saskatchewan, Canada S7N 5A8

Introduction

The rate of adoption for genetically modified (GM) crops at the close of the 20th century is arguably one of the most defining moments in the history of agriculture within that century. This rapid global adoption of GM varieties of oilseed rape, maize, cotton and soybeans was not without controversy. At the opening of the 21st century, the debate regarding new technologies shows no signs of abating. The number of contentious issues is wide and varied, ranging from social issues such as consumer acceptance to scientific issues such as gene flow.

Debate about gene flow has grown in importance over the past several years, especially with the commencement of crop trials involving pharmaceutical plants. There are two justifications for pursuing the technology of pharmaceutical plants. First, production of high-quality biological material presently is done using mammalian cells inside a bioreactor, which is very expensive and results in high drug costs that could potentially limit the number of people that benefit from new drugs. Secondly, even at the present time, there is an insufficient level of bioreactor capacity available to meet the current demand, let alone the expected increase in demand over the next decade.

The costs of the various antibody production systems which are valuable for the treatment of arthritis, herpes and cancer are widespread. The cost to produce 300 kg of active ingredient has been estimated for the following production systems: maize, US$8 million; eggs, US$12 million; tobacco, US$18 million; goats, US$57 million; and bioreactors, US$72 million (Pew Initiative on Food and Biotechnology, 2002). The prevailing technology of using mammalian cells to produce human antibodies costs in the range of US$105–175/g. It has been estimated (McCloskey, 2002) that transgenic plants might be able to produce the same amount of antibodies at a cost of US$15–190/g. The range of variation in anticipated plant-made pharmaceutical (PMP) costs arises from the prospect that the use of PMPs will lower production costs to a level that is economically feasible for potential new proteins that are presently prohibitively expensive to produce. Mammalian cell bioreactors take an average of 5–7 years to build at a cost of US$600 million, and McCloskey estimates that at present, 20–50 products could be delayed due to the unavailability of bioreactor capacity. The lack of production capacity is confirmed by the fact that production of four pharmaceutical products requiring biologics presently consumes 75% of mammalian cell fermentation capacity. By the end of this decade, there could be more than 80 competing antibody-dependent products with an estimated value that could exceed US$20 billion. The size of the market underlies the importance of exploring the potential of developing pharmaceutical plants.

As the economic importance of producing human antibodies grows and there are an increasing number of pharmaceutical plants grown commercially, the regulation of the plants will become a crucial issue. Of paramount importance will be assurances that the production of pharmaceutical plants will not co-mingle with conventional crop production destined for human consumption. The detection of drug proteins in processed food products could destroy the social trust in pharmaceutical crop technology and ultimately destroy the ability to take advantage of this technology.

The Importance of Innovative Human Therapeutics

The first proteins of human therapeutic value, interferon and insulin, were the first products of recombinant DNA technology to be produced in bacteria. The justification, at the time (mid-1970s), was that the fermentation and downstream processing technologies were mature, convenient and cost effective. The techniques of biotechnology were next used in the following transgenic products, first in animals in 1980 and for plants in 1983. These techniques of biotechnology had multiple applications, many of which at least initially had little direct relationship to clinical medicine. Although initially promising, the production of human therapeutic proteins such as antibodies in mammalian cells or animals (cows, goats, pigs and sheep) as yet has not had a wide adoption. Antibodies produced in mammalian expression systems are expensive and difficult to scale-up, and pose safety concerns due to potential contamination with pathogenic organisms or oncogenic DNA sequences. Plants on the other hand have become the principal focus for the production of antibodies, enzymes, vaccines and other therapeutic agents. With advances in transformation and expression systems, it can be expected that more pharmaceutically important genes from many species will be inserted into plants. Table 12.1 provides a comparison of the existing recombinant human protein production systems.

It is estimated that there are over 250,000 higher plant species, which are represented by a diversity of 300 different families. Natural products of botanical origin have included antimalarial

Table 12.1. Comparison of production systems for recombinant human pharmaceutical proteins.

System	Overall cost	Production time scale	Scale-up capacity	Product quality	Glycosylation	Contamination risks	Storage cost
Bacteria	Low	Short	High	Low	None	Endotoxins	Moderate
Yeast	Medium	Medium	High	Medium	Incorrect	Low risk	Moderate
Mammalian cell culture	High	Long	Very low	Very high	Correct	Viruses, prions and oncogenic DNA	Expensive
Transgenic animals	High	Very long	Low	Very high	Correct	Viruses, prions and oncogenic DNA	Expensive
Plant cell cultures	Medium	Medium	Medium	High	Minor differences	Low risk	Moderate
Transgenic plants	Very low	Long	Very high	High	Minor differences	Low risk	Inexpensive

Source: Ma *et al.* (2003).

drugs, aspirin, digitalis and the anticancer compound vinblastin, to name just a few. Overall, 25% of the prescription drugs in the US$300 billion pharmaceutical industry have plant origins (Cap *et al.*, 2002). Plants represent a cost-effective, convenient and safe alternative production system for several therapeutic vaccines, enzymes and recombinant antibodies (plantibodies). According to industry reports, these are the fastest growing class of new medicines for the treatment and prevention of human diseases such as Crohn's disease, rheumatoid arthritis, cancer and other potential ailments (e.g. inflammatory, central nervous system, cardiovascular and infectious diseases). Table 12.2 shows examples of the more than 40 recombinant proteins currently used in the treatment of human diseases or that are presently in the first or second phase of clinical trials.

There are ample demonstrations of plants and major food crops being efficiently transformed to produce, accumulate and store fully assembled and functional candidate transgenic products (Khachatourians *et al.*, 2002). Transgenic plants have become the alternative system for increasing the production capacity and for large-scale production and processing systems for antibodies (Faye *et al.*, 2003). The expression of immunotherapeutic proteins in plants has two major advantages compared with other expression systems: first, there is no risk of contamination with mammalian viruses or other pathogens; and, secondly, the production

of high amounts of antigens is cheap and, therefore, of great economic interest.

Although edible vaccines seem to be feasible, antigens of human pathogens have mostly been expressed in plants that are not attractive for human consumption (such as potatoes) unless they are cooked. Boiling may reduce the immunogenicity of many antigens. More recently, the technologies to transform fruit and vegetable plants have been perfected (Khachatourians *et al.*, 2002) to make plant products that can be eaten raw to pave the way towards edible vaccines. Delivery of immunotherapeutic pharmaceuticals in edible plant seeds is a legitimate source for oral vaccines, because antigens have been shown to be processed correctly in plants into forms that elicit immune responses when fed to animals or humans. Antigens expressed in maize are particularly attractive since they can be deposited in the maize seed and can be conveniently delivered to any organism that consumes grain. Edible vaccines can be used for active and passive immunization. Table 12.3 highlights some of the largely American-based firms that are presently conducting research using PMPs.

The cost of production of a new antibody requiring new manufacturing facilities is close to US$100 million and can take up to 4 years before being capable of production. The comparative cost of production of 1 kg of an antibody through a conventional bioreactor would require an

Table 12.2. Plant-based protein pharmaceuticals.

Proteins	Host plant	Clinical use	Reference
Anti-RhD IgG1 antibody	—	Alloimmunization of RhD– mothers with a RhD+ foetus	Bouquin *et al.* (2002)
Cholera toxin (CT)	—	—	Chikwamba *et al.* (2002)
Heat-labile enterotoxin(LT)	Maize	—	—
Hepatitis B surface antigen	Potato	Hepatitis	Richter *et al.* (2000)
Human interferon (IFN)-α	Potato	Listeria monocytogenes infection	Ohya *et al.* (2003)
Measles virus haemagglutinin	Carrot	Measles	Marquet *et al.* (2003)
Respiratory syncytial virus-F	Tomato	Respiratory infections, infancy/early childhood	Sandhu *et al.* (2000)
Spike protein (N-gS)	Potato	Swine TGEV	Gomez *et al.* (2000)

Table 12.3. Companies producing plant-based protein pharmaceuticals.

Company	Location	Proteins	Host plant
AltaGen Bioscience	Mirgan Hill, California	Antibodies	—
Dow AgroSciences Co.	Midland, Michigan	Proteins, antibodies	—
EPicyte Pharmaceuticals	San Diego, California	Antibodies	Maize
Exelixis Plant Sciences Inc.	San Francisco, California	Various proteins	Arabidopsis
Integrated Protein Technologies	St Louis, Missouri	Antibodies	Maize
Meristem Therapeutics	Clermont-Ferrand, France	Enzymes	—
Monsanto	St Louis, Missouri	Antibodies, hormones, proteins	Tobacco, maize
PhytaGenics	Richland, Washington	Plasma proteins, wound sealant	—
ProdiGene	College Station, Texas	Enzymes, vaccine	Maize
Scale Biology Corporation	Vacaville, California	Enzymes	—
Sunol Molecular Corp.	Miramar, Florida	Antibodies, anti-tissue factor	—

estimated US$3.75 million, whereas the same product generated in maize would cost US$33,000 (Hileman, 2002). Large-scale production of pharmaceutical grade antibodies requires further focus on development and operation of robust, reliable and cost-effective processes and their validation (Fahrner et al., 2001). Plant-produced therapeutic antibodies are already in clinical trials, but account for a very small percentage of the overall production from genetically engineered crops. Current estimates of transgenic food crops are nearly 60 Mha based on production in 2002.

PMPs are moving into the mainstream of biopharmaceutical manufacturing technologies. Daniell et al. (2001) review medical molecular farming, production of antibodies, biopharmaceuticals and edible vaccines in plants, arguing that they are cheap to produce and store, easy to scale-up for mass production and safer than those derived from animals. Here, we discuss recent developments in this field and possible environmental concerns.

Plants with Engineered Pharmaceutical Secondary Metabolite Products

Studies on plant secondary metabolites have been increasing over the last 50 years. These molecules are known to play a major role in the adaptation of plants to their environment, but also represent an important source of active pharmaceuticals. Plant cell culture technologies were introduced at the end of the 1960s as a possible tool for both studying and producing plant secondary metabolites. Despite all of the efforts of the last 30 years, plant biotechnologies have led to very few commercial successes for the production of valuable secondary compounds. Different strategies, using in vitro systems, have been studied extensively with the objective of improving the production of secondary plant compounds. Biotechnology has opened a new field, metabolomics, with the possibility of directly modifying the expression of genes related to biosynthetic pathways that lead to secondary plant metabolites and metabolic engineering. The delivery of PMPs could occur through ingestion of raw plant material or from a full or possibly partial conventional processing method.

Plants with Engineered Immunotherapeutic Products

Unlike the past, where undifferentiated cell cultures were the target for engineering of transgenic products, current interest and technologies have provided the opportunity to exercise the option for PMP production in leaves, tubers, fruits, hairy roots and other organs.

Leaves

Yusibov *et al.* (2002) showed immunogenicity of plant virus-based experimental rabies vaccine obtained in *Nicotiana benthamiana* and spinach (*Spinacia oleracea*) plants and used for parenteral immunization of mice. Mice immunized with recombinant virus were protected against challenge infection. Based on the previously demonstrated efficacy of this plant virus-based experimental rabies vaccine when orally administered in virus-infected unprocessed raw spinach leaves to mice, demonstrated significant antibody responses to either rabies virus. These findings provide a clear indication of the potential of the plant virus-based expression systems as a supplementary oral booster for rabies vaccinations.

Tubers

Microorganisms such as viruses, bacteria and certain protozoa cause diarrhoeic diseases of humans and animals. Gomez *et al.* (2000) have shown oral immunogenicity of the spike protein from swine-transmissible gastroenteritis coronavirus (TGEV) in potato tubers. Mice inoculated intraperitoneally or directly fed developed serum antibodies specific for TGEV-gS protein, demonstrating the oral immunogenicity of the plant-derived product. A different strategy for multivalent vaccine production from food plants for simultaneous protection against infectious virus and bacterial diseases was engineered by Yu and Langridge (2001). They created a fusion of cholera toxin (CT) B and A2 subunit with a rotavirus enterotoxin and enterotoxigenic *Escherichia coli* fimbrial antigen genes engineered into potato plants. Orally immunized mice indicated the presence of a strong immune response to the three plant-synthesized antigens. Diarrhoea symptoms were reduced in severity and duration in passively immunized mouse neonates following rotavirus challenge.

There are several programmes attempting to achieve global immunization for hepatitis B prevention and eradication. Oral immunization for hepatitis B virus with an 'edible vaccine' is an important strategy in this area. Richter *et al.* (2000) showed oral immunogenicity of recombinant hepatitis B surface antigen (HBsAg) in pre-clinical animal trials. Mice fed transgenic HBsAg potato tubers showed a primary immune response (increases in HBsAg-specific serum antibody) that could be greatly boosted by intraperitoneal delivery of a single subimmunogenic dose of commercial HBsAg vaccine, indicating that plants expressing HBsAg in edible tissues may be a new means for oral hepatitis B immunization. Kong *et al.* (2001) performed oral immunization with HBsAg expressed in transgenic potato plants. In fact, in transgenic plants, HBsAg accumulated intracellularly, suggesting natural bioencapsulation of the antigen as a means to provide protection from degradation in the digestive tract. These authors show transgenic plant HBsAg material to be superior in both inducing a primary immune response and priming the mice to respond to a subsequent parenteral injection of HBsAg. Marquet *et al.* (2003) used the desirability and palatability of edible vaccines containing antigens of human pathogens when they are expressed in crops that must be cooked. Boiling may reduce the immunogenicity of many antigens. They suggest that transformed carrot plants are a better choice. They were successful in the production of the immunodominant antigen of the measles virus. Their study may pave the way towards an edible vaccine against measles, which could be complementary to the current live-attenuated vaccine.

Seeds

Chikwamba *et al.* (2002) showed the ability of transgenic maize plants to produce enteropathogenic and diarrhoea-causing *E. coli* heat-labile enterotoxin (LT) and CT for use as an edible vaccine. Orally immunized mice had mucosal and systemic immune responses and survived challenge by oral administration of the diarrhoea-inducing toxins.

Fruits

Hepatitis B is a serious and debilitating disease, which can be prevented through vaccination. The cost and feasibility of widespread vaccination, however, are problematic. Gao *et al.* (2003) provides some theoretical and experimental directions for the production of large-scale, low-cost oral HBsAg vaccine using transgenic cherry tomatillo. Sojikul *et al.* (2003) engineered a signal peptide from soybean, a vegetative storage protein–HBsAg fusion protein, with enhanced stability and immunogenicity expressed in tobacco plant cells.

Sandhu *et al.* (2000) found that oral immunization of mice with transgenic tomato fruit expressing respiratory syncytial virus (RSV) fusion (F) protein induced a systemic immune response. Here a fruit-based edible subunit vaccine against RSV was developed by expressing the RSV F protein gene in transgenic tomato plants. The F gene was expressed in ripening tomato fruit under the control of the fruit-specific E8 promoter. Oral immunization of mice with ripe transgenic tomato fruits led to the induction of both serum and mucosal RSV-F-specific antibodies.

Plants with Engineered Human Pharmaceutical Enzyme Products

Leaves

A number of inborn errors in metabolism and enzyme deficiencies in humans can be corrected by the provision of enzymes. An example is Fabry's disease, a human lysosomal storage disorder. Human α-galactosidase is used to treat individuals with Fabry's disease. Unlike β-galactosidase which is found in many bacteria and fungi, α-galactosidase is scarce and its production very expensive. Tobacco plants engineered to produce this enzyme can yield 50 mg of enzyme/kg of freshly harvested tobacco plants (Potera, 2003). The US Food and Drug Administration (FDA) has granted Large Scale Biology Corp. of Vacaville, California orphan drugs status for the production of this therapeutic enzyme.

Seeds

ProdiGene Corporation is already in commercial scale-up of two proteins, the enzyme trypsin, and aprotinin, an inhibitor of protease from GM maize. Trypsin is used in the production of insulin, as an industrial protease and is also used as an oral medicine for wound care. Trypsin is usually produced from bovine pancreas. Aprotinin is used in cardiac surgery to prevent blood loss and in wound healing, and is obtained from bovine lungs.

PMP peptides

Bioactive peptides are 3–40 amino acids in length and carry a range of biological properties commercially useful to life and clinical sciences and agri-food industries. Areas of research and development of current interest are drugs active as hormones, insulin, anticoagulants, vaspopressins, brain neuropeptides and antimicrobial peptides. Peptide science has matured and in the years to come should have significant value to the pharmaceutical industry (Rocchi and Gobbo, 2002). The major method for large-scale manufacturing of peptides is still by chemical synthesis, or by fermentation of recombinant microorganisms (Verlander, 2002).

Tubers

Material costs and scale-up of plant-based peptide production argues favourably for the use of plant systems and transgenics. Ohya *et al.* (2003) have shown type I interferon (α/β IFN) production in transgenic potato. This is the first cytokine used for clinical applications against autoimmune, viral and neoplasmic diseases through oral administration.

Leaves

DeGray *et al.* (2001) showed expression of an antimicrobial peptide via the chloroplast genome to control phytopathogenic bacteria and fungi. Antimicrobial peptide MSI-99, an analogue of magainin 2, was expressed via the chloroplast genome to obtain high levels of expression in transgenic tobacco plants. Genetically engineering crop plants for disease resistance via the chloroplast genome instead of the nuclear genome is desirable to achieve high levels of expression and to prevent pollen-mediated escape of transgenes.

Challenges of Pharmaceutical Crop Production

The initial contentious issue regarding gene flow was oilseed rape (Smyth *et al.*, 2002), and that situation remains unchanged. Scientists and regulators are still in a conundrum at best or conflict at worst about the impacts and regulations of gene flow. The three leading types of crop used for pharmaceutical trials have been maize, tobacco and oilseed rape. The problem with these crops is that maize and rape are intended for human consumption and the potential for co-mingling or cross-pollination exists, raising concerns about the use of these crops for pharmaceutical trials. Table 12.4 presents the major transgenic crops, identifies whether they are used in pharmaceutical trials and examines the modality of consumption.

The leading transgenic cereal crop is maize where a total of 9% of global maize hectares are transgenic and grown in seven countries (James, 2003). The leading transgenic oilseed is oilseed rape, which is grown in Canada and the USA; transgenic rape accounts for 12% of global rape acres. Soybean, the sole transgenic pulse crop, is grown in seven countries and accounts for 51% of the global soybeans acres. The clear leading transgenic forage crop is tobacco, while a variety of transgenic fruits and vegetables are beginning commercial production.

Transgenic cereals, fruits and vegetables are, for the most part, consumed directly. The majority of cereals, such as maize and rice, are consumed directly. Transgenic fruits and vegetables are also consumed directly, while some processing would occur in the juice-making process. Transgenic oilseeds are largely used to produce processed oils, which are used to fry food. Most pulses are consumed directly, but the only transgenic pulse, soybean, is used largely for animal feed and in the production of tofu. Forage products are rarely consumed by humans, with a small amount of lucerne sprouts being the exception.

As the technology of pharmaceutical and transgenic plants rapidly moves from laboratory to field, the regulations developed to control these new crop varieties are being severely tested. While regulators in the USA have argued that the detection of ProdiGene's experimental pharmaceutical maize in a silo of soybeans late in 2002 is proof that the regulations are working (see below), the simple fact that a pharmaceutical crop that was supposed to be contained on-farm actually reached a grain terminal without being detected shows that the regulations are not stringent enough.

Table 12.4. Transgenic crops and human consumption.

Transgenic crop category	Specific transgenic crops	Use in pharmaceutical trials	Cross-pollination potential	Modality of consumption	
				Plant tissue(s) and organs[a]	Extracellular plant metabolic ingredients
Cereals	Maize, barley, rice, wheat	Yes	Low–medium	Direct	No
Oilseeds	Rape, flax, mustard, cotton, safflower	Yes	High	Mainly indirect	Yes
Pulses	Soybean	Yes	Low	Direct and indirect	No
Forage	Lucerne, clover, tobacco, sugarcane, sugarbeets	Yes	Medium	Very minimal indirect	No
Fruits and vegetables (including juices)	Poppy, cantaloupe, melon, radish, potato, squash, tomato, strawberry, lettuce and papaya	Yes	Low–high	Direct	Yes

[a]Plant cells, tissues and organs including rDNA or primary or secondary metabolites (i.e. excluding DNA such as oils, starches, proteins, amino acids and processed materials and tissues, including juice).

The containment of living plants is proving to be increasingly challenging given man's inability to control nature completely.

The issue of unintended gene flow became global news in the autumn of 2001 with the discovery that some varieties of Mexican maize contained transgenic material that should not have been there (Quist and Chapela, 2001). (This research is contested within the scientific community and was the subject of a NAFTA environmental review.) This topic continued to be important into the summer of 2002 as a research team led by Allison Snow of Ohio State University reported preliminary evidence suggesting that the trait from transgene insertions in sunflowers may be able to move to other plants, thus creating the conditions for 'superweeds' (Snow et al., 2003).

The problem multiplied in the autumn of 2002 and spring 2003 (Table 12.5). In November 2002, ProdiGene Inc. was fined US$250,000 for allowing experimental pharmaceutical maize volunteers to grow unintentionally to maturity within a soybean field. Inspectors with the United States Department of Agriculture's (USDA) Animal and Plant Health Inspection Services (APHIS) discovered the regulatory infringement. The resulting soybean crop was harvested and pooled in a commercial grain silo, thus contaminating an estimated 500,000 bushels of soybeans. The cost to ProdiGene for buying the contaminated soybeans and having them transported to be destroyed was an estimated US$3.5 million.

Within a month, the topic was once again making news headlines in North America. In mid-December, the US Environmental Protection Agency (EPA) fined Dow AgroSciences and Pioneer Hi-Bred for two separate regulatory violations. Dow was fined US$8800 for failing to meet all the defined conditions to prevent gene transfer with an experimental transgenic maize variety undergoing field trials at Molokai, Hawaii. The plot, which was one-tenth of an acre in size, failed to meet the EPA permit conditions because there was no windbreak made of wiliwili trees in place and the bordering rows (the outside 12–24 rows) were not of the variety specified in the permit. Pioneer was fined US$9900 for an experimental transgenic maize variety in Kauai that was planted in an unapproved location that turned out to be too close to other experimental maize varieties. The Pioneer permit from the EPA specified an isolation distance of 1260 feet, and this distance was not observed.

Finally, in April 2003, Dow was again fined for violating the EPA permit in Kauai. This time the fine was US$72,000 and resulted from the detection of 12 transgenic maize plants that contained an unapproved gene that is suspected to have come from the pollen from another experimental plot located nearby. Although Dow officials discovered this unplanned gene flow, Dow failed to notify the EPA promptly and EPA officials expressed disappointment over the delayed response by Dow. When this incident was reported in the *Washington Post* (April 24, 2003), the article stated that this incident was '. . . the latest setback for a biotechnology industry struggling to comply with government rules . . . some advocates say the problems cast doubt on a

Table 12.5. Impact of transgenic crop regulatory violations (2002–2003).

Company	Crop	Location	Violation	Impact
ProdiGene	Maize	Nebraska	Volunteer maize growing in soybean field	Fined US$250,000 and forced to pay clean-up costs of US$3.5 million
Dow AgroSciences	Maize	Hawaii	No tree windbreak and bordering rows	Fined US$8800
Pioneer Hi-Bred	Maize	Hawaii	Plot planted in unapproved location	Fined US$9900
Dow AgroSciences	Maize	Hawaii	Plants detected with unapproved gene and failure to notify the EPA promptly	Fined US$72,000

fundamental premise of government policy: that experimental varieties of corn or other crops can be planted in fields but kept out of food crops.'

Four separate, but related, regulatory violations within a 6 month period may be nothing more than a freak statistical occurrence that may never be observed again. More troubling, and probably more representative of the real issue, is that these regulatory violations could simply be the tip of the iceberg and that evidence of these regulatory violations could be documented continually for the foreseeable future.

Existing PMP Regulation

A closer examination of the regulatory systems in Canada and the USA reveals some surprising differences. Some would argue that the Canadian regulatory agencies have been more vigilant regarding transgenic crops than their US counterparts. Beginning in the early 1990s, Canadian regulators stated that all transgenic crops (as well as many mutagenic crops) would be treated as plants with novel traits (PNTs) and, therefore, subject to greater regulation than conventional crop varieties. Every new PNT requires mandatory oversight of their trials, efficacy and impact on safety of food, feed and the environment. Government agencies demand to see both the raw data and summaries of all tests performed, and have the final say on every introduction. The Canadian system also has a formal system of contract registration for risky industrial crops and imposes criminal penalties for infractions. While the Canadian regulators have not completed their development of special rules for PMPs, they have been very influential in directing companies away from areas deemed to be of higher risk (e.g. oilseed rape) by simply reminding the developers that such products are unlikely to be approved. Meanwhile, the Canadian Council for Biotechnology Information (CBI) is a smaller association than in the USA and has not developed the synergy that its counterpart, the Biotechnology Information Organization (BIO), enjoys in the USA. At least part of the reason is that the concentration of power in the Canadian government (through the Prime Ministerial structure, cabinet secrecy and party solidarity) limits the association's ability to gain access or find supporters within the government apparatus.

The initial regulations in the early 1990s in the USA were viewed by the industry as being too lax and therefore insufficient to establish trust with consumers. In response, the industry asked the regulators to strengthen the regulations for transgenic crops. Nevertheless, the American regulatory system has consistently been less rigorous in its approach to dealing with transgenic crops than regulators in Canada, e.g. most reviews are voluntary, non-transgenic novel traits are not reviewed and the regulatory agencies only see study summaries rather than raw data. As in Canada, the US regulators have not sorted out how to handle PMPs. The extra challenge they face is that they do not have the same powers and legal authority as Canadian regulators to direct developers away from crops. While the regulatory mechanism may be weaker, the industry association is considerably larger than in Canada and, given the more open nature of US governance, has better access and a stronger voice in the USA than in Canada. BIO is viewed by many as a very authoritative voice when speaking on issues affecting the industry.

On 6 March 2003, the APHIS division of USDA announced that they would strengthen mandatory permit conditions for field-testing transgenic crops, including field trials for PMPs (Animal and Plant Health Inspection Services, 2003). The number of site inspections will increase to five during the trial and two the following season. The permits for pharmaceutical trials will state that no maize can be grown within 1 mile (1.6 km) of the trial site and that no food or feed crop can be grown on the site the following season. The size of the buffer zone was doubled from 25 to 50 feet. This strengthening of regulatory requirements, in part, can be seen as a method to address the concerns that arose following the regulatory violations between 2001 and 2003.

Recent Industry–Government Regulatory Collaborations

In December 2003, the Center for Business Intelligence hosted a conference in Washington, DC titled, Plant-made Pharmaceuticals: Understand Regulations, Evaluate Costs/Benefits and Measure Risks. This was an attempt to bring several key regulators and representatives from leading firms together. Regulators from the FDA and APHIS provided a detailed outline of the regulatory structure as it applied to the production of PMPs. Included in this outline was an explanation of specific laws and acts that are applicable. The APHIS representative presented on the scope of the industry and how it declined dramatically between 2002 and 2003. In 2002, there were 20 permit approvals that totalled 134 acres (331 ha) of production, while in 2003, there were only 14 approvals totalling 25 acres (61.7 ha) of production. The various US regulatory agencies that have a mandate in the regulation of PMP production are beginning to coordinate their efforts regarding the regulation of PMP production. Representatives from the leading firms, public research institutions and concerned organizations outlined the processes they were using in the production of PMPs. One representative from Arizona State University highlighted some of the previous research results of vaccines for Norwalk virus and hepatitis B that were derived from PMP production (Mason *et al.*, 1996; Tacket *et al.*, 1998, 2000). It was asserted that one advantage of using food crops for the delivery of vaccines is that the human body is used to process this material. The speaker from the Union of Concerned Scientists provided a thorough review of some of the concerns that existed for both those involved in the PMP industry and society at large. This event was designed to be an information-sharing conference and, therefore, offered no opportunity to engage in detailed discussions about the actions needed to allow the industry to progress.

In contrast, in Canada, two events were organized in 2004 where interaction and dialogue were possible. The first event was a Technical Workshop organized by the Canadian Food Inspection Agency (CFIA) and held in Ottawa on 2–4 March. The title of this workshop was Technical Workshop on the Segregation and Handling of Potential Commercial Plant Molecular Farming Products and By-products. This workshop was organized to be the preliminary step in developing an effective and efficient regulatory framework for the commercial production of PMPs. This 3-day event offered presentations by experts within various aspects of the agriculture industry and provided the opportunity for numerous table discussions to identify key issues and future challenges facing the development of molecular farming. Participants included representatives from the PMP industry, Canadian and American federal regulators, agriculture organizations and experts in grain handling

and identity preservation. The workshop was divided into several sections for presentations and discussions, including: grain handling and transportation systems; identity preservation programmes; detection methods; gaps in knowledge and science; and code of best agriculture practices for PMP production. There was a general consensus among the participants that the bulk handling system is ill equipped to transport safely small amounts of novel products, such as PMP crop types. It was suggested, and well supported, that levels of regulation should be developed for PMP production so that the greater the level of risk associated with the specific protein being produced, the greater the regulatory requirements. Transportation from the production site to the point of processing was identified as a way of reducing human error in the movement of the harvested crop. The PMP industry would be well advised to use identity preservation systems to ensure that there is a closed loop production system. Testing to exceedingly low levels of tolerance continues to be a challenge given specific information on the novel protein being produced is protected due to confidential business information. Additionally, testing cannot provide an absolute guarantee of purity. One important gap in the knowledge of PMPs that was raised was the issue of what is needed regarding plant material from PMP crops that remains in the field following harvesting. More information on this topic is needed prior to the development of a functional regulatory framework. Participants jointly assisted in the development of a preliminary outline for a code of best agricultural practices for PMP production. Some key next steps were identified at the end of the workshop. First, efforts will be increased regarding Canadian–American regulatory harmonization in the area of PMP production. Representatives from the CFIA and USDA APHIS expressed the opinion that it was very important for both parties to ensure that dialogue continues to take place regularly between both organizations. Secondly, representatives from the CFIA and Health Canada pledged to work towards finalizing a memorandum of understanding between the two regulatory bodies that outlines each organization's respective regulatory responsibilities. Thirdly, the CFIA and Health Canada will continue to work in harmony to develop an appropriate regulatory framework that allows for the successful development of a commercial PMP industry in Canada.

The second event was held on 26 and 27 April and was co-hosted in Ottawa by the Canadian Agri-Food Research Council (CARC) and the Genomics, Ethics, Economics, Environment, Law and Society (GE[3]LS) Project within Genome Prairie. The event was called the Biobased Molecular Production Systems Workshop 2004. The objective of the 2-day workshop was to move towards the development of a policy framework leading to an action plan for responsible commercialization of products of biobased molecular production systems.[1] Invited participants and speakers included academics, leading industry representatives, government regulators and non-governmental organizations. Presentations from the four federal departments (Agriculture and Agri-Food Canada, CFIA, Health Canada and Industry Canada) that have a role in the regulation of molecular products provided a thorough overview of the structure and scope of the existing regulation regime. Following the regulatory perspective was a session that focused on the social context of molecular farming. Presentations in this area examined European regulations and ethics, consumer perspectives, social issues in Canada, patent challenges, producer perspectives, food processor and manufacturer perspectives and patient group perspectives. Concluding the presentations was a series of talks that focused on industry experiences. Presentations were heard that covered pharmaceutical applications, industrial product applications, animal pharmaceutical opportunities and the industry's code of conduct.

Several sessions of table discussions were held with the intention that the discussions would be building toward a supportive environment for products of biobased molecular production systems. Numerous themes emerged from the table discussions, with the leading ones being commitment to science-based regulations, communication and promotion of biobased molecular products, understanding consumer perceptions and market reactions, and regulatory harmonization and transparency.

This workshop was held to develop an action plan to assist the industry in moving forward. As the workshop closed, three key strategies were identified for the action plan. The first was the agreement from the industry representatives to establish a CEO level industry group. This industry group was asked to identify common concerns and challenges and to ensure that the correct message was conveyed to the correct people and decision makers. The second achievement was the agreement from the government regulators to establish an interdepartmental working group. This group would be tasked with the goal of working towards creating a cohesive and coordinated approach to regulating products of biobased molecular production systems. The third strategy was the decision to establish a permanent advocacy link between the industry and the elected government. A senior representative of the biotechnology industry was approached to serve as the advocate with the Prime Minister's Office Chief Science Advisor. Participants encouraged these three key strategies to work closely together and asked that regular lines of communication be developed between the three to ensure that issues and opportunities are addressed promptly.

Events, conferences and workshops of the preceding nature will be key to the advancement of the plant-made pharmaceutical industry. Continued interaction between industry and government allows for the development of key contacts which facilitates the lines of communication that will lead to the development of a regulatory pathway. The development of such a pathway will be of benefit to both industry and government as it will provide a framework of consistent and transparent regulatory procedures. At the end of the day, this is really what both sides are asking for, a clear and understandable system of regulations.

Conclusions

The challenge of PMPs is going to be to structure a fully integrated regulatory system that effectively evaluates, manages and communicates about the risks of a system, and ultimately one that both enforces and is seen to enforce failures. In spite of the US regulatory changes, there is an apparent inability of regulators to enforce the regulations. In the ProdiGene case, the cost of the fine, clean up and destroying the contaminated soybeans was estimated to be US$3.75 million. The problem with imposing such a large cost on a small biotechnology company is that there is seldom enough cash flow within the company to pay a fine of this magnitude. The American government had to lend ProdiGene the money to pay the fine. This is symptomatic of the biotechnology industry as a whole, as small biotechnology companies do not have sufficient financial resources to pay large regulatory violation fines. The problem is that if firms know that governments will provide loans or loan guarantees in the event of fines from regulatory violation, then there is less incentive for firms to adopt standards that improve the control of pharmaceutical crops. If existing enforcement mechanisms are found wanting or are lacking, trust will be hard to sustain.

Part of the future solutions to the conundrum of PMPs will be stricter regulation. Options such as: (i) a closed loop system to confine PMPs to select authorized seed processors, certified growers, manufacturing sites and pharmaceutical firms; or (ii) production of PMPs in regions where there are no major food crops are being examined by regulatory systems. It is encouraging to observe that industry–government dialogue and discussions relating to the development of a regulatory framework have been established and are continuing. While large-scale commercial PMP production may not occur in the immediate future, the development of clear, consistent and efficient regulations needs to continue moving forward. As the technology relating to molecular farming evolves and improves, additional issues will arise and, for this reason, the establishment of a coordinated approach to the development of regulations will be a crucial determining factor in the successful growth of this industry.

A continued focus on science-based regulations will be of major importance for this industry as several environmental organizations have already publicly expressed concern about the production and commercialization of PMP products. This has been a consistent demand from firms in the PMP industry and, in all of the presentations by both Canadian and American regulators, there has been no mention of incorporating additional non-scientific regulations. While there are possible transparency concerns (i.e. the location of trials not being public knowledge), concerned organizations need to recognize that firms need to be able to protect the basis of their innovative research; however, firms and regulators need to be as open as possible.

It should be noted that the future of PMP trials in the state of Hawaii is uncertain due to the regulatory violations. Environmental groups are lobbying to prevent further PMP field trials. Continued regulatory violations may remove production options for firms in the PMP industry.

Continued lack of production areas or use of feed crops may reduce the value of PMPs as expression vectors and, ultimately, raise the cost of the drug beyond many potential users' ability to pay. This would reduce the benefits of PMPs and things may change for the worse, where costs are translated into more severe health effects and death.

Increasing the regulatory harmonization between Canada and the USA should be an objective of regulators in both countries. Several of the firms involved in this industry are presently conducting field trials in both countries, and the regulatory agencies need to be aware that draconian regulations may ultimately drive the firms in this industry to the other jurisdiction. The regulation of an industry as innovative as PMPs will continue to be a challenging issue for the short term, but the collaborative efforts between both industry and government and between governments provides a level of confidence that the commercialization of PMP products will take place under an efficient and effective regulatory regime.

Note

[1] Biobased molecular production systems is meant to include both plant and animal molecular production systems.

References

Animal and Plant Health Inspection Services (2003) Field Test Releases in the US Retrieved from the World Wide Web at: http://www.nbiap.vt.edu/cfdocs/fieldtests1.cfm

Bouquin, T., Thomsen, M., Nielsen, L.K., Green, T.H., Mundy, J. and Dziegiel, M.H. (2002) Human anti-Rhesus D IgG1 antibody produced in transgenic plants. *Transgenic Research* 11, 115–122.

Cap Gemini, Ernst and Young (2002–2003) *World Pharmaceutical Frontiers*, p. 77.

Chikwamba, R.J., Cunnick, D., Hathaway, J., McMurray, H., Mason, H.S. and Wang, K. (2002) A functional antigen in a practical crop: LT-B producing maize protects mice against *Escherichia coli* heat labile enterotoxin (LT) and cholera toxin (CT). *Transgenic Research* 11, 479–493.

Daniell, H., Streatfield, S.J. and Wycoff, K. (2001) Medical molecular farming: production of antibodies, biopharmaceuticals and edible vaccines in plants. *Trends in Plant Science* 6, 219–226.

Degray G., Rajasekaran, K., Smith, F., Sanford, J. and Daniell, H. (2001) Expression of an antimicrobial peptide via the chloroplast genome to control phytopathogenic bacteria and fungi. *Plant Physiology* 127, 852–862.

Fahrner, R.L., Knudsen, H.L., Basey, C.D., Galan, W., Feuerhelm, D., Vanderlaan, M. and Blank, G.S. (2001) Industrial purification of pharmaceutical antibodies: development, operation, and validation of chromatography process. *Biotechnology and Genetic Engineering Reviews* 18, 301–327.

Faye, L., Lerouge, P. and Gomord, V. (2003). Production de molecules pharmaceutiques. *Annales de Pharmacologie* 61, 109–118.

Gao, Y., Ma, Y., Li, M., Cheng, T., Li, S.W., Zhang, J. and Xia, N.S. (2003) The oral vaccination of HBV by cherry tomatillo transformed with HBsAg. *Chinese Journal of Virology* 19, 31–35.

Gomez, N., Wigdorovitz, A., Castanon, S., Gil, F., Ordas, R., Borca, M.V. and Escribano, J.M. (2000) Oral immunogenicity of the plant derived spike protein from swine-transmissible gastroenteritis coronavirus. *Archives for Virology* 145, 1725–1732.

Hileman, B. (2002) Drugs from plants stir debate. *Chemical Engineering News* 80, 22–25.

James, C. (2003) *Global Status of Commercialized Transgenic Crops: 2002. ISAAA Brief No. 27.* Retrieved from the World Wide Web at: www.isaaa.org

Khachatourians, G.G., McHughen, A., Nip, W.K., Scorza, R. and Hui, Y.H. (eds) (2002) *Transgenic Plants and Crops.* Marcel Dekker Inc., New York.

Kong, Q., Richter, L., Yang, Y.F., Arntzen, C.J., Mason, H.S. and Thanavala, Y. (2001) Oral immunization with hepatitis B surface antigen expressed in transgenic plants. *Proceedings of the National Academy of Sciences of the USA* 98, 11539–11544.

Ma, J.K-C., Drake, P.M.W. and Christou, P. (2003) The production of recombinant pharmaceutical proteins in plants. *Nature Genetics* 4, 794–805.

Marquet, B.E., Bouche, F.B., Steinmetz, A. and Muller, C.P. (2003) Neutralizing immunogenicity of transgenic carrot (*Daucus carota* L.) derived measles virus hemagglutinin. *Plant Molecular Biology* 51, 459–469.

Mason, H.S., Ball, J.M., Shi, J., Jiang, X., Estes, M.K. and Arntzen, C.J. (1996) Expression of Norwalk virus capsid protein in transgenic tobacco and potato and its oral immunogenicity in mice. *Proceedings of the National Academy of Sciences of the USA* 93, 5335–5340.

McCloskey, R. (2002) Presentation to the Pew Initiative on Food and Biotechnology Conferance. Retrieved from the World Wide Web at: http://pewagbiotech.org/events/0717/ConferenceReport.pdf

Ohya, K., Matsumura, T., Itchoda, N., Ohashi, K., Onuma, M., Sugimoto, C. and Vasil, I.K. (2003) Protective effect of orally administered human interferon (HuIFN)-alpha against systemic *Listeria monocytogenes* infection and a practical advantage of HuIFN-alpha derived from transgenic potato plant. *Plant Biotechnology 2002 and Beyond.* Proceedings of the 10th IAPTC and B Congress, Orlando, Florida. Kluwer Academic Publishers, Dordrecht, The Netherlands, pp. 389–391.

Pew Initiative on Food and Biotechnology (2002) Pharming the field: a look at the benefits and risks of bioengineering plants to produce pharmaceuticals. Retrieved from the World Wide Web at: http://pewagbiotech.org/events/0717/ConferenceReport.pdf

Potera, C. (2003) TMV uses could convert tobacco into a weed for health. *American Society for Microbiology News* 69, 376.

Quist, D. and Chapela, I. (2001) Transgenic DNA introgressed into traditional maize landraces in Oaxaca, Mexico. *Nature* 414, 541–543.

Richter, L.J., Thanavala, Y., Arntzen C.J. and Mason, H.S. (2000) Production of hepatitis B surface antigen in transgenic plants for oral immunization. *Nature Biotechnology* 18, 1167–1171.

Rocci, R. and Gobbo, M. (2002) Peptide science in the years to come. *Chimica Oggi* 20, 26–29.

Sandhu, J.S., Krasnyanski, S.F., Domier, L.L., Korban, S., Osadjan, M.D. and Buetow, D.E. (2000) Oral immunization of mice with transgenic tomato fruit expressing respiratory syncytial virus-F protein induces a systemic immune response. *Transgenic Research* 9, 127–135.

Smyth, S., Khachatourians, G.G. and Phillips, P.W.B. (2002) Liabilities and economics of transgenic crops. *Nature Biotechnology* 20, 537–541.

Snow, A., Pilson, D., Rieseberg, L., Paulsen, M., Pleskac, N., Reagon, M., Wolf, D. and Selbo, S. (2003) A Bt transgene reduces herbivory and enhances fecundity in wild sunflowers. *Ecological Applications* 13, 279–286.

Sojikul, P., Buehner, N. and Mason, H.S. (2003) A plant signal peptide–hepatitis B surface antigen fusion protein with enhanced stability and immunogenicity expressed in plant cells.

Proceedings of the National Academy of Sciences of the USA 100, 2209–2214.

Tacket, C.O., Mason, H.S., Losonsky, G., Estes, M.M., Levine, M.M. and Arntzen, C.J. (2000) Human immune responses to a novel Norwalk virus vaccine delivered in transgenic potatoes. *Journal of Infectious Diseases* 182, 302–305.

Tacket, C.O., Mason, H.S., Losonsky, G., Clements, J.D., Levine, M.M. and Arntzen, C.J. (1998) Immunogenicity in humans of a recombinant bacterial antigen delivered in a transgenic potato. *Nature Medicine* 4, 607–609.

Verlander, M. (2002) Large scale manufacturing methods for peptides – a status report. *Chimica Oggi 20, 62–66.*

Washington Post (2003) Firm fined for spread of altered corn genes. Thursday, April 24, E4.

Yu, J. and Langridge, W.H.R. (2001) A plant-based multicomponent vaccine protects mice from enteric diseases. *Nature Biotechnology* 19, 548–552.

Yusibov, V., Hooper, D.C. Spitsin, S.V., Fleysh, N., Kean, R.B., Mikheeva, T., Deka, D., Karasev, A., Cox, S., Randall, J. and Koprowsk, H. (2002) Expression in plants and immunogenicity of plant virus-based experimental rabies vaccine. *Vaccine* 20, 3155–3164.

13 Are GURTs Needed to Remedy Intellectual Property Failures and Environmental Problems with GM Crops?

Geoff Budd

General Council, Grains Research and Development Corporation, Canberra, Australia

Introduction

Proponents enthuse that genetic use restriction technologies (GURTs) 'have the potential to benefit farmers and others in all size, economic and geographical areas' (Collins, 2003). First, they provide seed companies and plant breeders with stronger control over plant varieties, by preventing growers reproducing seed. This enables higher cost-recovery, so provides more incentives for plant breeding. Secondly, they may offer significant environmental benefits, overcoming a lot of the concerns about genetically modified (GM) crops.

Opponents condemn GURTs as 'an immoral application of agricultural biotechnology' (Action Group on Erosion, Technology and Concentration, 2002b) leading to bioserfdom (Action Group on Erosion, Technology and Concentration, 2002a). In particular, they attack the loss of farmers' ability to save and re-use seed, and raise environmental concerns.

This chapter examines these claims, in relation to two types of GURTs, variety GURTs (V-GURTs) and trait-specific GURTs (T-GURTs). It examines weaknesses in plant breeders' rights (PBRs), and patent and contract laws, but questions whether breeders really need the extra control over how their plant varieties are used that the GURTs are intended to provide. The chapter also examines some of the environmental impacts of these GURTs.

These issues are examined from the Australian perspective of a small, developed economy reliant on import of germplasm and export of agricultural products, with plant breeders seeking a sustainable funding model.

Characteristics of GURTs

GURTs are tools of modern technology that regulate gene expression. GURTs could introduce a genetic switch mechanism to prevent unauthorized use of either particular plant germplasm or traits associated with that germplasm. Alternatively, GURTs could have environmental or safety purposes. The end result is, by definition, a GM plant and will be regulated as such.

These technologies could have many potential applications. They are still under development, and are some time away from field testing. The discussion in this chapter must be considered in the context that the technologies, and the claims made about them, are at a very early and therefore uncertain stage of development.

This chapter discusses the two most commonly debated forms of GURTs. Other forms of GURTs are not discussed.

Variety-level GURTs (V-GURTs)

V-GURTs aim to render the next generation (such as seed) sterile, so have become known as 'terminator technology'. Growers cannot save and replant the seed.

Several ways of creating V-GURTs have already been patented but not put into practice. The most well known patent was issued in March 1998, to Delta and Pine Land Co. and the United States Department of Agriculture (US patent No. 5,723,765). One of the methods claimed uses a lethal gene and promoter, separated by a blocking sequence that allows normal embryo development, producing first-generation seed. When sold, the seed is treated with a chemical inducer that removes the blocking sequence. The lethal gene expresses and the next generation of seed will be sterile (Food and Agriculture Organization, 2001a).

In the method patented by Zeneca (US Patent No. 5,808,034), the lethal gene expresses by default. The plant breeder applies a chemical in all generations, which blocks expression of the lethal gene. When ready to sell the seed, the breeder stops applying the chemical, making the next generation sterile.

V-GURTs are somewhat similar to hybridization. F_1 hybrids are seeds of two different parents that have been conventionally cross-bred in one generation, to incorporate a desired trait. In the next generation, diversification (out-crossing) occurs and the seed will not fully replicate the desired trait. Productivity declines rapidly with each generation (Food and Agriculture Organization, 2001a). Hybridization can, so far, generally only be used commercially with out-crossing crops, such as oilseed rape (canola), maize and sorghum (Goeschl, 2003).

In-breeding (selfing) crops, such as wheat, barley, cotton and soybeans, replicate well in subsequent generations. Growers can save seed (subject to any legal restrictions) and subsequent generations will maintain the productivity of the original seed. V-GURTs have the most potential for in-breeding crops, by removing growers' ability to save seed.

Trait-specific GURTs (T-GURTs)

With a T-GURT, only the desired transgene trait is switched on or off, rather than rendering the seed sterile or ending subsequent reproduction of the germline. T-GURTs would be useful in all crops, including hybrids. This could enable a variety to incorporate, for example, tolerance to salt or heavy metals. Growers would themselves decide whether

to activate or keep active the trait, at the time of buying the seed, or later by purchasing the activator from the technology provider (Jefferson, 1999). Syngenta has lodged a US patent application for use of chemicals to control gene expression (US Patent application US200110022004A1). For example, growers in a particular area where the trait was needed could activate the trait using a proprietary chemical, supplied by the plant breeder's vertically integrated chemical company.

State of the Technology

In 2001, the Food and Agriculture Organization (FAO) estimated that 'the pace of biotechnology development should allow GURTs and their products to become functional in the next five to ten years' (Food and Agriculture Organization, 2001a). By May 2003, Syngenta held eight US patents on V-GURT technology, and others held a total of nine patents (Action Group on Erosion, Technology and Concentration, 2003b). There has been little subsequent public evidence of rapid progress – the 2001 estimate might have been optimistic.

The Delta and Pine Land (D&PL) patent provoked an international furore, largely because of the (then) proposal for Monsanto to buy D&PL, which did not proceed. The debate has continued at an international level, despite some of the patent owners stating that they did not intend to commercialize the technologies. See, for example, Monsanto CEO Robert Shapiro's open letter to Rockefeller Foundation President Gordon Conway, dated 4 October 1999. It is questionable whether these early technologies will ever be used commercially in seed, because they appear to be too complex and not reliable enough. However, they are useful research tools, and more robust methods are being developed (Jefferson, 1999), although commercial release does not seem likely in the near future.

The Australian Context

Changing nature of breeders and breeding

Traditionally, plant breeding in Australia has been largely carried out in the state Departments of Agriculture and universities, particularly for the major crops of wheat and barley. Each breeding programme has been small and narrowly focused. That is rapidly changing, as public funding is withdrawn, or public funds are invested in a more commercial manner.

The Grains Research and Development Corporation, a major funder of grains research and development in Australia, is encouraging consolidation of breeding programmes, to obtain a smaller number of programmes with sustainable scale and to eliminate duplication. Large private industry players are becoming involved, as AWB Limited and Syngenta have established the Longreach wheat breeding joint venture. However, the breeding programmes in most state departments have uncertain futures. The Western Australia Department of Agriculture has announced a restructuring of plant breeding into a research institute with less overall funding. Two other states have recently announced major restructures of their agriculture departments, and it is not known what the effect will be on plant breeding.

Another key factor has been the changing nature of plant breeding itself. Gene technology is taking an increasingly important role, both in the research tools used and in the varieties produced. These technologies can dramatically increase the speed of current breeding strategies. Instead of taking 10–12 years to develop a conventional (non-GM) plant variety, it can now take 6–8 years (interview with Greg Fraser, Director, Enterprise Grains Australia, 22 May 2004).

The new technologies also allow breeders to use different 'pathways', using a broader array of germplasm than previously and giving them fundamentally new capabilities. This latter approach ultimately offers great benefits to all users of the germplasm, but will cost a lot for its developer (Donnenwirth *et al.*, 2004). Most of the Australian patents relating to use of gene technology are owned by overseas companies. It is, therefore, increasingly important that plant breeders be able to both access technology to stay competitive and recover their costs of breeding.

These changes in Australia are no different from changes taking place overseas (Knight, 2003). However, several factors exacerbate the difficulties faced by Australian plant breeders. First, Australia's markets are relatively small for a plant breeder seeking to recover the costs of developing a new plant variety. These costs will often be much higher for a GM variety, particularly once the high costs of meeting Australia's

stringent regulations on GM crops are taken into account.

GM cotton and oilseed rape have been approved for public release in Australia, by the Office of the Gene Technology Regulator (OGTR) (see list of licences at http://www.ogtr.gov.au/gmorec/ir/htm). However, key state governments have imposed moratoria preventing general release of grains used for food (not including cotton). These moratoria are not likely to be removed in the short term. Monsanto has recently announced it is suspending research into GM rape in Australia, as part of its worldwide refocusing of GM research, but Bayer insists it is proceeding with its investment in GM rape (White, 2004). At a time when Australia is still searching for a sustainable plant breeding model, there is controversy about whether scarce public and private resources should be invested in breeding GM crops.

Also, in Australia, most germplasm is imported and then locally adapted. Over 90% of wheat varieties have been developed from International Maize and Wheat Improvement Centre (CIMMYT) germplasm (Brennan, 1998). This makes it imperative that Australian breeders continue to be able to access elite germplasm from overseas, at an affordable cost.

These factors make Australian breeders and ultimately growers especially vulnerable if the balance of incentives between breeders and others is not right.

Farm-saved Seed and Royalties

International organizations have attacked V-GURTs as being unethical and immoral, leading to bioserfdom (Action Group on Erosion, Technology and Concentration, 2002b), alleging that they will prevent growers in the developing world from saving seed, thus making them vulnerable to multinational seed companies. The International Seed Federation disputes this (Collins, 2003). The issue is linked by several non-government organizations (NGOs) to highly emotional claims about threats to food security, biodiversity and inability to access intellectual property. It is portrayed largely in developed versus developing country terms.

The right to farm-saved seed is largely an economic issue in Australia, as in most developed economies. Most germplasm is imported and there is little conservation of landraces, which is a key international justification for farm-saved seed. Like growers anywhere, Australian growers are constantly looking for the most profitable and sustainable farming systems. This involves choosing the best combination of seed varieties (both for each crop and for subsequent rotations), herbicides and crop management regimes. The cost of seed is only a relevant factor in deciding what varieties to grow, and other major factors include the cost of fertilizers, herbicides and insecticides, labour and land. These are balanced against likely returns for sale of the crop, once royalties are deducted. Growers also seek to maximize yield by appropriate crop rotation and tools such as precision agriculture.

However, much of Australia's land used for cropping is more marginal than, for example, that used in Canada and the USA. This makes crops less reliable in much of Australia. Growers are reluctant to pay high prices for seed, because of the risk that they may get a poor crop. Past experience of plant breeders and seed companies has been that for conventionally bred varieties offering only incremental advances, seed-based royalties have not been successful at recovering breeding costs (AWB, 2002).

It remains to be seen whether varieties incorporating V-GURTs or T-GURTs could provide a sufficiently large advance that growers are prepared to pay substantial seed royalties. It may be that for V-GURTs, plant breeders using a seed royalty may only be able to charge a relatively small premium compared with the benefit the V-GURT offers. Unless the variety offers a large agronomic advantage, breeders may be better off using the emerging system of end-point royalty backed by contract – which raises many of the enforcement problems that V-GURTs are intended to avoid.

Despite initial hostility to end-point royalties (usually collected at the point of receipt of the harvested crop), growers are becoming much more prepared to pay an end-point royalty than an equivalent seed royalty. This is because they are only paying in proportion to the success of their crop, so share the risk with the breeder. However, end-point royalties present real enforcement problems for plant breeders. These are discussed below in relation to PBR and patent laws.

Similar issues arise with T-GURTs. Jefferson's suggestion, of relying on charging for use of the activator chemical (Jefferson, 1999), has possibilities

where the plant breeder and chemical company are linked or come to an agreement. It would probably be relatively attractive to growers, because the grower only pays where he wants the benefit of the trait.

Control Over Unauthorized Use – Has Intellectual Property Law Failed Breeders?

V-GURTs as the 'silver bullet' solving breeders' problems

V-GURTs' proponents argue that intellectual property laws and contracts provide seed companies and plant breeders with too little protection over how their seed varieties are used. They allege that there is considerable 'leakage' of revenue, because growers are easily able to save or trade seed, and to avoid paying royalties. Monsanto, for example, estimated that it could lose up to 25% of royalty payments in the USA (Smyth *et al.*, 2002).

It is expensive to enforce intellectual property rights, and other breeders are able quickly to appropriate the benefits of new varieties. The situation is analogous to piracy of CDs and DVDs – it is easy to cheat, and hard for the intellectual property owners to police.

V-GURTs, like anti-copying technology for copyright material, offer intellectual property owners a 'silver bullet'. If a grower simply cannot re-use seed, intellectual property laws and contracts become largely irrelevant, although patent law helps protect the V-GURT itself. As the USDA and D&PL stated in announcing their March 1998 V-GURT patent described above:

> The principal application of the technology will be to control unauthorised planting of seed of proprietary varieties (sometimes called 'brown-bagging') by making such practices non-economic since unauthorised saved seed will not germinate, and would be useless for planting (Action Group on Erosion, Technology and Concentration, 2002b).

V-GURT proponents argue that despite the high cost of developing the V-GURT and protected trait, transaction and enforcement costs are significantly reduced if contracts and intellectual property laws are not needed (Food and Agriculture Organization, 2001a; Smyth *et al.*, 2002).

This section examines the above claims. It concludes that although V-GURTs' proponents overstate the extent of the problems with intellectual property and contract laws, there are significant issues to be resolved. However, opponents of V-GURTs significantly overstate the issues V-GURTs raise about the balance of rights between the current plant breeder and others.

Plant Breeders' Rights (PBRs)

Scope of coverage

The Australian *Plant Breeders Rights Act* 1994 (PBR Act) is based on the 1991 revision of the International Convention for the Protection of New Varieties of Plants (UPOV Convention, 1991). The PBR Act applies to plant varieties, which includes a plant grouping 'that can be considered as a functional unit because of its suitability for being propagated unchanged'. To be registrable, a plant variety must be distinct, uniform and stable. PBRs do not apply to the individual genes, tools and products of genetic engineering, and so provide quite narrow coverage.

PBRs also extend to any subsequent plant variety that is 'essentially derived' from the initial variety (sections 4 and 12), and aim to prevent 'copycat' activity. They may provide powerful protection for a breeder who has inserted a gene into a plant variety, by preventing another plant breeder making very slight changes to the variety and then seeking to exploit it. The essential derivation test remains largely untested at a formal level, both in Australia and overseas (Plant Breeder's Rights Office, 2002; Le Buanec, 2004).

The UPOV Convention (1991) lists 'transformation by genetic engineering' as an example of an essentially derived variety (Article 14(5)(c)). However, the PBR Act does not include this example in the definition of an essentially derived variety. The Report of the Australian Expert Panel on Breeding states that 'Genetic modification, whether done by 'traditional' or 'biotech' methods, is not necessarily 'copying'' (Plant Breeders' Rights Office, 2002). Janis and Kesan (2002) take the same approach, on the basis that 'only minimal modification . . . puts one outside the scope of protection. . .'. Le Buanec, Secretary-General of the International Seed Federation, is highly critical of the Australian approach,

arguing that genetic engineering was the trigger point for introducing the concept of essentially derived variety into UPOV (Le Buanec, 2004).

Exceptions

PBRs are subject to several limitations and defences, much more so than patents. The two key exceptions are breeding (section 16) and farm-saved seed (section 17(1)).

Breeding exception

Under the breeding exception, any act done for non-commercial or experimental purposes, or for breeding other plant varieties, does not infringe PBRs. Tools such as molecular markers have dramatically shortened conventional breeding times. A breeder used to have 10–12 years from the time it released its variety to the time a competitor could release an improved variety based on it. Now it may have as little as 6–8 years (interview with Greg Fraser, Director, Enterprise Grains Australia, 22 May 2004). Breeders now obtain a much shorter period of protection of a PBR-protected variety than was the case when the UPOV Convention was last modified in 1991.

In the case of a GM variety, Ewens suggests the breeding exception means that PBR:

> does not provide adequate protection for a breeder who has inserted a gene into a new plant variety because it allows another breeder to purchase the genetically altered, and PVPA [the US PBR equivalent] protected, plant and breed the new gene into a new variety (Ewens, 2000).

The Report of the Australian Expert Panel on Breeding discusses the complexities in developing a market-ready GM variety (Plant Breeders' Rights Office, 2002). Taking that GM variety, and using it to develop a new variety that is not just an essentially derived variety, will also take several years of time and considerable effort. It may take as little as 5 or 6 years' research (interview with Richard Brettell, Program Manager, GRDC 23 May 2004), which may not be enough time to recover the high cost of investment in a GM variety and regulatory approval costs.

However, the second breeder will then need to obtain approvals from the OGTR. If the second breeder can rely on the very detailed information provided by the first breeder in relation to the transformation event in the first variety, its regulatory costs should be relatively low in obtaining OGTR approval. However, the OGTR is required to maintain confidentiality over commercial-in-confidence information supplied to it by an applicant. The OGTR probably would not be allowed to use information supplied by one breeder in approving a competing modified variety. The second breeder would probably be required to obtain and provide its own data, adding 2 years and close to $1 million to the approval process (interview with Richard Brettell, Program Manager, GRDC 23 May 2004), and maybe much more.

This is much less than the 20 years protection provided by patents (see below), but still gives the original breeder 6 years or more to exploit the gene or variety. In that time, it should be able to build up a strong market presence, including using trademarks to build a brand (see Blakeney, 2002). If well marketed, this advantage should continue for some years after a competitor has brought an improved variety to market.

If this ultimately proves not to be enough time for breeders to make a return on investment, PBR protection could be strengthened by revising the breeders' exception so that it does not apply for 'x' years (see Donnenwirth *et al.* (2004) and discussion in Le Buanec (2004)). This would need to be done at the UPOV level – Australia could not do so unilaterally or it would become inconsistent with UPOV. Such a change should be handled cautiously as the breeders' exception is a fundamental part of the UPOV system.

Farm-saved seed

The farm-saved seed exception allows growers to save and replant seed, but not to trade it with other growers. It was first made explicit in the UPOV Convention (1991), and is optional, so Australia could remove it or weaken it. For example, in the European Union (EU), farmers (except small farmers) can save seed, but must pay a royalty of 50% of the 'normal' licence fee to the PBR holder (European Regulation on the Protection of Plant Varieties 1994 and 1998 Implementing Rule). As discussed above, in Australia, the justification for the farm-saved seed exception is largely economic.

Whether the farm-saved seed exception helps or hinders plant breeders depends largely on the royalty system used. For a breeder or seed company relying on a seed-based royalty, farm-saved seed reduces the amount of seed the grower will need to buy in future years, and so reduces the overall royalties the breeder or seed company will make during the life of the plant variety. This in turn reduces the incentive for the breeder to invest.

In contrast, a breeder relying on an end-point royalty wants the variety to be adopted as widely and quickly as possible, to maximize tonnage of the variety harvested. Farm-saved seed would encourage rapid adoption, increasing the royalties payable to the breeder. The risk for a breeder is that at the time of delivery, the grower may mis-declare the plant variety as being another variety subject to a lower or no end-point royalty, in order to minimize or avoid paying the royalty. Technology capable of providing an accurate, cost-effective and time-sensitive 'on the spot' determination of a plant variety is likely to be needed. It is under development, but is not yet available.

The breeder needs to maximize its return and be compensated for the cost of germplasm development. If the breeder (the PBR owner) has no contractual relationship with the grower, it is difficult to establish when a grower is likely to be avoiding paying the end-point royalty and to obtain the chain of evidence and intent. It must rely largely on tip-offs that a grower is mis-declaring deliveries of a variety to start investigations. It must then incur the substantial cost of suing for breach. It can recover damages or an account of profits (PBR Act section 56(3)), but these are unlikely to be significant from an individual grower. The author is personally aware that there have been many breaches of PBRs in Australia. To date, there have not been any successful legal actions for breach of PBRs in the grains industry.

Using contracts in conjunction with PBRs

Using grower contracts in conjunction with PBRs addresses a lot of the enforcement problems described above, without needing GURTs. On their own, contracts provide weak protection for plant breeders, because the requirement for privity of contract means it is only enforceable between the parties. The plant breeder will have difficulty suing a third party who obtains the seed from the contracted grower. However, a contract can require the grower to keep records, making it much easier to monitor the contracted grower's conduct and use of the seed. The breeder can have a good idea of how much seed is likely to be produced, and so of how much is likely to be sold (and subject to royalty) or kept for re-sowing.

Preventing grower-to-grower trading (brown bagging) is also crucial in ensuring the breeder can track what happens with the seed. A significant discrepancy in these amounts may indicate the grower is breaching their contract and/or PBR Act obligations, triggering an investigation. This is particularly important where an end-point royalty is used, rather than a seed-based royalty.

It is much easier to sue for breach of contract than for breach of PBRs alone. A well-drafted damages clause in a contract also makes it easier to obtain significant compensation than for breach of legislation. The contract also acts as a powerful deterrent. A grower who has signed a contract with (for example) AWB Seeds, knows that AWB Seeds will easily be able to access his property and check his records, under the contract. This is much more 'real' than the risk of a patent or PBR owner tracking him down and trying to establish a breach of the patent or PBRs.

Contracts with growers are a relatively new development in Australia. In the past, public sector plant breeders have depended on support from growers for their political survival and have not been prepared to do anything likely to alienate growers. This is rapidly changing. Private plant breeders are much more willing to impose onerous contracts on growers and to charge higher royalties, where the variety offers a demonstrable advantage. Public breeders are also becoming more confident, so grower contracts are becoming common.

One concern is that so far there are no common forms or terms of contract; each company has a very different approach. Growers risk being confused by a multitude of different contracts, and breeders are incurring substantial administrative costs. If plant breeders were able to agree on an industry-standard model of grower contracts (whilst taking care not to breach trade practices prohibitions against anti-competitive conduct), growers and breeders could both benefit.

Key industry bodies (seed companies, the Australian Seed Federation and the Grains Research and Development Corporation) are currently considering strategies to remedy some of the problems with PBR, outlined above. These steps include common forms of contract and record-keeping requirements of growers. If successful, these steps will increase the confidence of breeders in the PBR and contract system. If they fail, V-GURTs and T-GURTs may yet be an attractive alternative.

Patents

Before the development of modern biotechnology, patents were not available to protect plant varieties. Conventional plant breeding was not considered to involve an inventive step, and new varieties were obvious, not inventive. However:

> with the extension of patent protection to recombinant methods for producing transgenic plants and the resulting products, patents have begun to assume an increasing significance in plant variety protection. The broader ambit of patent rights is a particular advantage of this form of intellectual property protection, covering . . . plants, seeds and enabling technologies (Blakeney, 2002).

This has caused controversy about whether patents relating to genetic material provide patent holders with rights that are too strong at the expense of others. V-GURTs will exacerbate this debate, because a plant variety protected by a V-GURT would not need to meet the threshold requirements for patentability, and the exceptions to patents will not apply.

Patentability and scope of coverage

Under the *Patents Act* 1990 (Patents Act), plant varieties and processes relating to plants are clearly patentable in Australia, as each of the threshold requirements has been held to be satisfied. Australia, along with the USA and Japan, now allows greater patenting of plant varieties than does the EU and most other jurisdictions (Donnenwirth *et al.*, 2004).

The key development in patenting of genetic material was the 1980 *Diamond v Chakrabarty* decision in the US Supreme Court. It held that gene sequences that have been isolated and purified are inventions, not mere discoveries. In *Kirin-Amgen Inc.* (1995), the Australian Deputy Commissioner of Patents took a similar view. He decided that a claim to the naturally occurring DNA sequence for a naturally occurring protein would only be a discovery, but a claim to an isolated and purified DNA sequence was patentable subject matter (Ludlow, 1999).

In *Grain Pool* (2000), the Australian High Court relied on *National Research Development Corp* (1959) and the line of US decisions from *Chakrabarty* to hold that there is no intrinsic impediment to the patentability in Australia of plant varieties, and that patenting of plant varieties could coexist with PBRs.

In *J.E.M Ag Supply v Pioneer Hi-Bred International, Inc.*, the United States Supreme Court held that plants can be protected by patents, despite being protectable by PBRs legislation. Janis believes that the *Pioneer Hi-Bred* decision is likely to lead to more plant breeders lodging patent applications in the USA than previously (Janis, 2002). The same is happening in Australia, with several breeding programmes currently preparing or considering patent applications for plant varieties.

Limitations and exceptions

There are few relevant limitations or exceptions to a patent holder's rights to exclude others from exploiting the patent. The two potential limitations are a research exemption and compulsory licensing. In particular, there is no farm-saved seed exception. If a grower retains seed containing patented cells and grows a further crop without a licence from the patent owner, he will be in breach of the patent. This has not been tested in Australia, but has been litigated in both the USA and Canada. The controversial Canadian case *Monsanto v Schmeiser*, in which the Supreme Court of Canada narrowly upheld that Percy Schmeiser had breached Monsanto's patent, is the best known example.

Compulsory licences

Compulsory licences can be granted where the reasonable requirements of the public have not been satisfied (Patents Act sections 133 and 135). However, this is a narrower ground than in the USA, so compulsory licensing is rarely used in Australia. The requirement to seek a court order also makes it cumbersome and expensive to apply, particularly in deciding on the appropriate royalty

to be paid to the patent holder (Lawson, 2002). It seems to serve mainly as a negotiating tool (Nicol, 2001).

Research exemption

There is no explicit research exception in Australia. Any research exception under present Australian law is likely to be very narrow (Smith, 2003), certainly much narrower than the breeding exception to PBRs. In the USA, the 2003 Federal Circuit of Appeals decision in *Madey v Duke University* has effectively eliminated the research exception. A given use would have to be 'solely for amusement, to satisfy idle curiosity, or for strictly philosophical inquiry.' There is a real possibility that *Madey* could be followed in Australia.

A plant breeder is clearly able to prevent others incorporating the patented product in their own plant variety and commercializing it during the life of the patent, unlike under PBRs. This crucial difference between patents and PBR reduces the risk of another plant breeder quickly piggybacking on the patent owner's work and so increases the incentive to conduct plant breeding, making plant breeders seek patents in preference to PBR where possible. However, does the lack of a research exception give too much control to the patent owner? In the absence of a clear research exception, gene patents, particularly on research tools, have been criticized as potentially hampering research (Heller, 1998). Obtaining licences can be time consuming and costly, or the patent holder may impose stringent conditions. This may increasingly be the case as patent owners use patents to deny their competitors access to the critical enabling technologies (Jefferson, 1999).

It is not yet known whether patenting of biotechnology blocks research, or whether the concern is a transitional phenomenon (Heller, 1998). In 2004, the Australian Law Reform Commission acknowledged the deterimental effect of the current uncertainty in Australian law. It recommended that the Australian law be amended, to clarify that certain (limited) experimental use of a patented invention does not infringe patent rights (Australian Law Reform Commission, 2004). At worst, the current lack of a research exception may mean that, even without GURTs, technology owners may have too much protection, not too little as implied by GURTs' proponents. GURTs may make this situation worse, as discussed below.

Enforcement against growers still has problems

Patents, when available, provide plant breeders with much stronger protection against growers avoiding paying royalties than do PBRs. However, they are still difficult and expensive to enforce against growers.

In seeking to enforce the patent against growers, the plant breeder will encounter many of the practical problems faced by an owner of PBRs, described above. The evidentiary issues are easier than for PBRs, because there is no farm-saved seed exception. As with PBRs, use of grower contracts greatly reduces these enforcement problems, but the damages awarded against each grower would probably be small. The main benefit of a successful suit for patent infringement would be its deterrent effect. Patent litigation is also horrendously expensive, although this would be more of a problem for growers than plant breeders. It may also lead to the patent itself being challenged, as this is a common response to claims of patent infringement.

Balance of Rights

PBR, possibly with modification to the breeders' exception, used in conjunction with grower contracts, appears to provide breeders with a high enough level of protection and incentive to invest, particularly for incremental breeding – but only just. This is balanced by the exceptions enabling research to continue and maintaining growers' freedom to keep seed. Patents, which are useful for more fundamental changes, tilt the balance slightly in favour of the patent owner, largely because of the longer period of protection and lack of research or breeding exception.

Effect of V-GURTs on Research

GURTs' potential to swing the balance of rights further towards the GURTs' owner also concerns opponents. Jefferson warns in relation to V-GURTs that:

> Such biologically-based protection systems, that would effectively remove the policy control of governments of intellectual property rights in plant varieties, may be broader (entire genome, any seed), more effective (100% control) and less

limited by time (compared to patents and licences) (Jefferson, 1999).

V-GURTs will be added to a plant variety together with some desirable trait (such as salt tolerance) that gives significant agronomic advantage. The role of the V-GURT is to protect against unauthorized use of all of the genetic material, but particularly the desirable trait (Food and Agriculture Organization, 2001a).

If V-GURTs can prevent access to germplasm, they will have serious implications for research. A plant breeder builds on many generations of earlier effort (Watson, 1997). PBRs allow all of the previous and new work to be used for further research. A patent only protects the new work, although the narrow scope (or lack) of the research exception may make it difficult for breeders to build on that new work during the life of the patent.

In isolation, this is not a major concern, because there are other, slightly older, varieties available. However, in the long term, this could become a real problem if V-GURTs prevent access to germplasm and become widely used. The FAO surmises that even if V-GURTs promoted increased investment in a crop in the short term, ultimately plant breeders simply may not be able to access the range of elite germplasm needed to improve varieties incrementally, more than is the case now (Food and Agriculture Organization, 2001a).

The V-GURT itself can often be protected by patent. This minimizes the risk of another plant breeder reverse-engineering it. Standard patent law applies – other researchers can then seek a licence to use the V-GURT, or risk using it without a licence. Once the patent expires, they can use the V-GURT at will.

V-GURTs could limit researchers' ability to access the desired trait for further research, in two ways. First, they could stop a plant variety from reproducing, making it harder to use in research. Secondly, they could enable the owner of the plant variety to protect it as a trade secret, avoiding the need to disclose the details of the plant variety and its desirable traits.

Does No Reproduction Mean No Research?

Whether or not the sterility of a V-GURT will prevent researchers using the seed in research, to use and build on the desired trait(s) will depend on the location of the V-GURT (interview with Doug Waterhouse, Registrar, Plant Breeder's Rights Office, 18 August 2003). If the location of the V-GURT prevents reproduction of the germplasm containing the desired trait, it will be very difficult for researchers to use the seed to obtain the desired trait; otherwise it will not prevent researchers doing so.

Jefferson suggests that a V-GURT could block 'the further exploitation not only of the trait or traits he/she intends to protect, but of the whole associated genome.' However, the V-GURT 'would not prevent by itself the imitation of a certain product by other companies or entities that may possess the technical capabilities to reverse-engineer or otherwise duplicate the 'technically protected' seed' (Jefferson, 1999).

In January 2003, a memorandum prepared by the Office of UPOV concluded that a GURT 'prevents access to germplasm, hampers research and breeding progress and sustainability, and limits benefits to society'. 'Plant material of varieties containing GURTs cannot be used as genetic material for further breeding; free access to genetic resources will be hindered by GURTs' (UPOV, 2003a).

The USA and others strongly attacked the UPOV statement. The USA stated that 'research can still be conducted on the GURTs variety through asexual reproduction, single or multiple cell cultivation, etc.' (US Patent and Trademark Office, 2003). The UPOV Office subsequently withdrew most of its claims (UPOV, 2003b). The Action Group on Erosion, Technology and Concentration (ETC) claimed that 'UPOV has succumbed to the strong-arm tactics of the US government' in withdrawing the memorandum (Action Group on Erosion, Technology and Concentration, 2003a). However, in reality, the 10 January memorandum was a poor-quality document, issued without appropriate consultation, so the criticism was warranted.

In discussing this issue with several researchers, the author received contradictory responses; clearly the USA is right in stating that 'GURTs technology is so new that there are very little reliable, relevant, scientific, economic and social data.'

By comparison, there is less risk of T-GURTs significantly hampering research, because they will not prevent reproduction of the seed. The T-GURTs will be more closely linked to the desired trait, which should make it easier to identify and use the desired trait in research.

No Disclosure may Hamper Research

Both patents and PBRs require the applicant to disclose details of the invention or plant variety for which protection is sought; the obligation to disclose is the cornerstone of patent and PBRs protection (Blakeney, 2002).

Even where a V-GURT does not prevent research taking place, it offers the potential for plant breeders to protect the identity and genealogy of a commercially successful line as a trade secret, in the same way as has been done for the inbred parents of hybrid maize. Researchers would have to try to reverse-engineer the plant variety by hit and miss. This would make it more difficult and time consuming to use the desired trait to produce further varieties.

For hybrids, it is possible to 'chase the selfs', by sifting through a bag of the hybrid seed to find self-pollinated seed of the hybrid's parents that may have been accidentally included, thus identifying the parents. This makes trade secrets a vulnerable form of protection for hybrids (Janis, 2002). It is unclear whether V-GURTs would be vulnerable in similar ways.

If it becomes clear that V-GURTs will hinder or prevent research, it may be necessary to require disclosure of information about the plant variety, perhaps in the same way as required for PBRs, to enable other breeders to use the trait. This could be imposed by the OGTR, as a condition for release, under the *Gene Technology Act* 2000. Alternatively, the plant breeder could be required to give others licences to use the V-GURT and trait. Presumably it would be necessary to require plant breeders using the trait and V-GURT to pay reasonable compensation.

Market Consolidation and Competition

These issues are debated in many industries. Jefferson notes that internationally:

> the current structure of the biotechnologically sophisticated component of the seeds industry

already displays high levels of concentration – either through ownership or through contractual relationships – and the trend seems to be accelerating (Jefferson, 1999).

He worries that if GM crops dominate, and access to biotechnology innovations is restricted, then:

> the possibility exists for market dominance by a few suppliers with potentially serious consequences for technology choice and price fixing (Jefferson, 1999).

The FAO acknowledges these issues, noting that it would be an incremental process. It concludes that:

> Whether GURTs raise concern over the development of possible monopoly power in the sector will depend partly on the extent to which firms or other entrants can develop competing or alternative products, with or without their own GURT technologies (Food and Agriculture Organization, 2001a).

Australia holds very little intellectual property in relation to GURTs, so plant breeders looking to use GURTs would need to obtain them under licence. This also raises the issue of whether GURTs hamper research use of key traits, discussed above.

Normally this would not be a concern; for all major crops there are a number of competing plant varieties, with none offering such a significant advantage over the others that they cannot compete. In that situation, in theory at least, it would be in the GURT owner's interests to license it as widely as possible, to increase the number of varieties in which it is used. This would particularly apply for a T-GURT that used a chemical to activate it; the owner of the chemical would want it to be used as widely as possible, in a managed way (Jefferson, 1999).

However, that will change from time to time; for example, rust resistance in wheat varieties could break down and a new strain of wheat be developed with significantly better rust resistance. It could then dominate the market – especially if protected by a GURT. Unlike PBRs, which require the owner to ensure reasonable public access to the variety (PBR Act section 19), there would be little that could be done to control how the GURT owner exploited the variety, unless its conduct was blatant.

Environmental Issues

There are concerns that GM crops could cause several environmental problems, such as spreading by pollen blowing into other crops or by horizontal gene transfer. As a preliminary point, the GM crops released in several countries to date have been deemed to be 'substantially equivalent' to non-GM crops, and regulation has been minimal. Future generations of GM crops are likely to offer health or nutritional benefits, then industrial or pharmaceutical properties. These may require much more effective gene control systems than have been achieved to date (Smyth *et al.*, 2002).

Proponents of GURTs claim they could reduce or eliminate a lot of the claimed environmental concerns about GM crops (whilst denying the validity of those concerns in the first place). On the other hand, GURTs are claimed to cause other environmental problems. The ETC Group thunder that 'The current campaign to promote seed sterility as an environmentally beneficial technology is illogical and dangerous' (Action Group on Erosion, Technology and Concentration, 2002b). ETC provides little evidence to support this statement.

A fundamental issue is whether V-GURTs are likely to be reliable enough. In its 2004 report 'Biological Confinement of Genetically Engineered Organisms', the United States National Research Council highlights that the technology is at an early stage of development and that there is a real risk of incomplete sterility. A prerequisite for V-GURTs offering environmental advantages will be that they are in fact reasonably reliable.

Environmental Benefits of GURTs

V-GURTs could potentially contain the spread of genes from GM crops, reducing or removing a claimed key problem with GM crops (Food and Agriculture Organization, 2001a; Collins, 2003). Genes from GM crops can in theory spread in several ways. These include through volunteer seed and gene transfer into other related species (wild relations), or co-mingling of seed. The potential liability issues may be immense, at all levels of the food chain, although it is unclear whether or not the risk of being successfully sued is significant. In the recent Starlink litigation in the USA, GM maize that had not been approved for human consumption ended up in crops and food for human consumption (Jones, 2002; Nelson, 2002). Aventis, the owner of Starlink, has reportedly budgeted over US$1 billion for compensation and other costs (Smyth *et al.*, 2002). Reducing these risks would increase the chances of market acceptance of GM crops, particularly future generations of crops.

Volunteers

GURTs will not prevent co-mingling of seed once it has been harvested. However, it can reduce the extent to which volunteers are a problem. There are three categories of volunteer seed. First, it can be seed that did not germinate in the first year's crop, but remains dormant in the ground until later years. This occurs in several crops (Dr Ray Hare, Senior Research Scientist, NSW Department of Agriculture, 5 August 2003), but not major crops including maize, soybeans, cotton, wheat or rice. V-GURTs will not prevent this happening, so the later crops planted in the same field will contain a small (and declining) proportion of the GM seed. However, that seed will itself be sterile, so the problem is minimal.

Secondly, volunteer seed can also be seed that is left on the ground during harvesting. It is inevitable that some seed is spilt during the harvesting process. However, with a V-GURT, that seed will be sterile, so will not contaminate the next crop.

Thirdly, pollen from open-pollinating crops such as oilseed rape and maize can blow from one crop to another. V-GURTs will ensure that wind-blown pollen will be sterile, so will not contaminate the next harvest.

Superweeds

By reducing the presence of the GM crop in subsequent years, the V-GURT also reduces the risk of creating 'superweeds' (there have been no such superweeds to date, so this is a theoretical risk). For GM crops such as Roundup® Ready oilseed rape or InVigor® oilseed rape, growers must spray the field in the following years to remove volunteers (this is also common practice for non-GM crops). Although unlikely, this may create resistance to the herbicide, if it is not used strictly in accordance

with instructions. In the case of InVigor® rape, the OGTR has decided this risk is negligible, because the herbicide is not widely used in Australia (Office of the Gene Technology Regulator, 2003). Monsanto, in its application for Roundup® Ready rape, proposed a detailed farm management plan to reduce this risk.

However, some plant breeders are sceptical of whether such plans can be effective in the medium or long term. Spraying is expensive and complicated (Smyth *et al.*, 2002). Many growers will not follow the spraying programme properly, deliberately or otherwise. Some breeders forecast development of resistance as being inevitable.

Gene out-crossing

Another risk is that of transferring genes from the GM crop to other crops or organisms (outcrossing), particularly where wild relatives exist locally. It is only a risk for open-pollinators such as oilseed rape and maize that release abundant pollen; for self-pollinating crops such as wheat or soybeans, the risk is very low (Pendleton, 2004). In Australia, where the germplasm for the key crops is almost entirely imported, this is unlikely to be an issue, with the potential exception of rape. In any case, the V-GURT can minimize this risk (Lamkey, 2002), because it is assumed that the V-GURT would probably transfer along with the desired gene (Jefferson, 1999). The FAO comments that V-GURTs 'might be justified' in this situation (Food and Agriculture Organization, 2001b).

Other advantages

GURTs could also provide agronomic advantages. V-GURTs could prevent pre-harvest sprouting, which is a real problem in Australia. T-GURTs could be used to switch a desired trait on or off in favourable or unfavourable situations, such as drought (Food and Agriculture Organization, 2001a).

In order to minimize the above risks, growers are often required to maintain buffer zones around the GM crop (Smyth *et al.*, 2002). This could make a GM crop uneconomic. V-GURTs can reduce or avoid the need for buffer zones.

Environmental Risks of GURTs

Drops in yield

The FAO claims that V-GURTs could lead to the 'possibility of pollination of neighbours with GURTs pollen, leading to yield drops in cultivated areas'. However, 'the probability may be low, given the multiple gene recombination events that would need to accompany outcrossing' (Food and Agriculture Organization, 2001a). In Australia, this would not be a significant issue.

Food security

One long-term issue that cannot be ignored, albeit a remote possibility, is that:

> The greatest potential risk to food security associated with wide adoption of V-GURTs may be the increased dependence on seed production and distribution by a few commercial suppliers and the vulnerability of such supply to disruption, either civil or environmental (Jefferson, 1999).

Biodiversity

V-GURTs potentially threaten biodiversity, by displacing locally adapted genetic material (Food and Agriculture Organization, 2001a; Visser *et al.*, 2001). This entails increased vulnerability to unexpected environmental stress and could result in increased food insecurity (Jefferson, 1999). However, the same could be said of increasing concentrations of conventionally bred varieties, so 'in high intensity farming systems the impact may be minor' (Food and Agriculture Organization, 2001a). As discussed above, in Australian crops, most germplasm is imported and then locally adapted to cope with Australia's unique conditions such as poor soils and variable climate. This argument is of little relevance in Australia.

Jefferson argues that T-GURTs can have a positive effect on genetic diversity by adding 'platform' value to varieties bred for local conditions (Jefferson, 1999). This could be useful in Australia, because it would encourage tailoring for local conditions.

Summary of Environmental Issues

Overall it appears that if GURTs work as effectively as promised, they offer real solutions to make GM crops safer in the environment, without the need for restrictions on their use. However, the technology is at such an early stage that 'guessing at the probabilities of such outcomes – whether positive or negative in effect – in the absence of working examples of V-GURTs (or indeed T-GURTs) is not likely to be helpful.' (Jefferson, 1999).

GURTs could reduce the need for restrictions on GM crops, making GM technology more acceptable to the community, and enabling growers and plant breeders to benefit from GM technology. However, because V-GURTs were originally stated as aiming to prevent unauthorized use, it may now be hard to convince the public of V-GURTs' and T-GURTs' claimed environmental benefits, over the campaigns of certain NGOs lobbying against them.

Conclusion

It is still too early to know whether the claims made for or against V-GURTs and T-GURTs are correct. It is clear that many of the arguments from both their proponents and opponents are quite exaggerated, particularly those of the opponents. However, the intellectual property, economic and environmental issues raised in relation to the potential use of V-GURTs and T-GURTs may be critical for Australia as it attempts to sustain scientific research underpinning a viable plant breeding industry. At this stage, intellectual property and contract laws appear to provide plant breeders with just enough protection and incentive – although there are still many problems to be overcome. If those problems cannot be overcome, V-GURTs and T-GURTs may yet be useful to enable breeders to make enough of a return to justify investment in plant breeding.

In addition, the environmental benefits offered by V-GURTs and T-GURTs may reduce or remove many of the potential emerging environmental issues with GM crops, making them more acceptable to growers and consumers.

References

Action Group on Erosion, Technology and Concentration (2002a) Defend food sovereignty: terminate Terminator. January 2002, www.etcgroup.org

Action Group on Erosion, Technology and Concentration (2002b) ETC group responds to Purdue University's recent efforts to promote seed sterilization – or Terminator – as an environmental protection technology', 1 May 2002, www.etcgroup.org

Action Group on Erosion, Technology and Concentration (2003a) Who calls the shots at UPOV? 17 April 2003, www.etcgroup.org

Action Group on Erosion, Technology and Concentration (2003b) Terminator technology – five years later. *Communiqué*, Issue 79, May/June 2003, www.etcgroup.org

Australian Law Reform Commission (2004) *Genes and Ingenuity*. Report No. 99, pp. 317–345. http://www.alrc.gov.au/inquiries/title/alrc99/index/html

AWB (2002) AWB Limited Submission to Australian Competition and Consumer Commission in support of notification under section 93(1) of the *Trade Practices Act* 1974 (Cth), section 2.2, 3 September, 2002.

Blakeney, M. (2002) Intellectual property, biological diversity, and agricultural research in Australia. *Australian Journal of Agricultural Research* 53(2), 127–147.

Brennan, J and Fox, P. (1998) Impact of CIMMYT varieties on the genetic diversity of wheat in Australia, 1973–1993. *Australian Journal of Agricultural Research* 49, 175–178.

Collins, H.B and Krueger, R.W. (2003) *Potential Impact of GURTs on Smallholder Farmers, Indigenous and Local Communities and Farmers Rights*. Paper made available to CBD's Ad Hoc Technical Expert Group on the Impact of GURTs on Smallholder Farmers, Indigenous & Local Communities and Farmers Rights, February 19–21, 2003.

Donnenwirth, J., Grace, J. and Smith, S. (2004) Intellectual property rights, patents, plant variety protection and contracts: a perspective from the private sector. *IP Strategy Today* No. 9-2004, 19–34.

Eaton, D. and Tongeren, F. (2002) *Genetic Use Restriction Technologies (GURTs): Potential Economic Impacts at National and International Levels*. Report 6.02.01. Agricultural Economics Research Institute (LEI), The Hague, Netherlands.

Ewens, L. (2000) Seed wars: biotechnology, intellectual property, and the quest for high yield seeds. *Boston College International and Comparative Law Review* 23, 285, at footnote 60.

Food and Agriculture Organization (2001a) *Food and Agriculture Organisation, Commission on Genetic Resources for Food and Agriculture, Potential Impacts of Genetic Use Restriction Technologies (GURTs) on Agricultural Biodiversity and Agricultural Production Systems*. Rome, 2–4 July, 2001.

Food and Agriculture Organization (2001b) *Food and Agriculture Organisation, Report of the Panel of Eminent Experts on Ethics in Food and Agriculture*. Rome, 26–28 September, 2001.

Goeschl, T. and Swanson, T. (2003) The development impact of genetic use restriction technologies: a forecast based on the hybrid crop experience. *Environment and Development Economics* 8, 149–165.

Heller, M. and Eisenberg, R. (1998) Can patents deter innovation? The anticommons in biomedical research. *Science* 280, 698–701.

Janis, M. (2002) *Intellectual Property Issues in Plant Breeding and Plant Biotechnology*. Biotechnology gene flow, and intellectual property rights: an agriculture summit. 13 September 2002, Purdue University, Indianapolis.

Janis, M. and Kesan (2002) US Plant variety protection: sound and fury . . .? *Houston Law Review* 39, 727–778

Jefferson, R.A. (1999) *Technical Assessment of the Set of New Technologies Which Sterilise or Reduce the Agronomic Value of Second Generation Seed, as Exemplified by US Patent No. 5,723,765 and WO 94/03619*. Expert paper prepared for CBD Secretariat, 30 April 1999.

Jones, A. (2002) What liability of growing genetically engineered crops? *Drake Journal of Agricultural Law* Autumn, 621.

Knight, J. (2003) A dying breed. *Nature* 421, 568.

Lamkey, K. (2002) *GMOs and Gene Flow: a Plant Breeding Perspective*. Biotechnology gene flow, and intellectual property rights: an agriculture summit. 13 September 2002, Purdue University, Indianapolis.

Lawson, C. (2002) Patenting genes and gene sequences and competition: patenting at the expense of competition. *Federal Law Review* 30, 115.

Le Buanec, B. (2004) Protection of plant-related innovations: evolution and current discussion. *IP Strategy Today* No. 9-2004, 1–18.

Ludlow, K. (1999) Genetically modified organisms and their products as patentable subject-matter in Australia. *European Intellectual Property Review* 21, 298–312.

Nelson, A. (2002) Legal liability in the wake of Starlink™: who pays in the end? *Drake Journal of Agricultural Law* Spring, 241.

Nicol, D. and Nielsen, J. (2001) The Australian medical biotechnology industry and access to intellectual property: issues for patent law development. *Sydney Law Review* September 2001, 347–376.

Office of the Gene Technology Regulator (2003) *Commercial Release of Genetically Modified (InVigor® Hybrid) Canola, Risk Assessment and Risk Management Plan*. DIR 021/2002, Canberra, 25 July 2003.

Pendleton, C. (2004) The peculiar case of 'Terminator' technology: agricultural biotechnology and intellectual property protection at the crossroads of the third green revolution. *Biotechnology Law Report* 1, 1–29.

Plant Breeder's Rights Office (2002) *Clarification of Plant Breeding Issues Under the Plant Breeder's Rights Act 1994*. Report of the expert panel on breeding, www.affa.gov.au

Smith, C. (2003) Experimental use exception to patent infringement – where does Australia stand? *Intellectual Property Forum* 53, 14.

Smyth, S., Khachatourians, G. and Phillips, P. (2002) Liabilities and economics of transgenic crops. *Nature Biotechnology* 20, 537.

United States National Research Council (2004) *Biological Confinement of Genetically Engineered Organisms*. National Academies Press, Washington (see www.nap.edu/catsalog/10880.html).

UPOV Convention (1991) *The 1991 Revision of the International Convention for the Protection of New Plant Varieties of Plants*. http://www.upov.int/en/publication/conventions/1991/act1991.htm

UPOV (2003a) Memorandum prepared by the office of UPOV on the Genetic Use Restriction Technologies. Geneva, January 10, 2003.

UPOV (2003b) Position of UPOV concerning Decision VI/5 of the conference of the parties to the CBD, 11 April 2003.

US Patent application US200110022004A1, Goss *et al.*/Syngenta (2001) Control of gene expression in plants by receptor mediated transactivation in the presence of a chemical ligand. Filed 21 March 2001.

US Patent No. 5,808,034, Bridges, I. *et al.*/Zeneca Limited (1998) Plant gene construct comprising male flower specific promoters. Granted 15 September 1998.

US Patent No. 5,723,765, Oliver *et al.*/Delta Pine Land Co. (1998) Control of plant gene expression. Granted 3 March 1998.

US Patent and Trademark Office (2003) *Proposal of the United States of America Regarding Procedural and Substantive Issues on the GURTs Memorandum Submitted by UPOV to the CBD*. 28 March 2003.

Visser, B., Van der Meer, I., Louwaars, N, Beekwilder, J. and Eaton, D. (2001) The impact of 'terminator' technology. *Biotechnology and Development Monitor* 48, 9–12.

Watson, A. (1997) *The Impact of Plant Breeder's Rights and Royalties on Investment in Public and Private Breeding and Commercialisation of Grain Cultivars*. Paper prepared for the Grains Research and Development Corporation, March 1997.

White, S. (2004) Bayer sticking to GM guns in Australia. *The Land* 19 May 2004.

14 Economic Effects of Producing or Banning GM Crops

Janine Wronka and P. Michael Schmitz

Department of Agricultural Policy and Market Research, University of Giessen, Diezstrasse 15, D-35390 Giessen, Germany

Introduction

The incidence of food scares in the European Union (EU)[1] has risen in recent years. Food scandals such as bovine spongiform encephalitis (BSE), dioxin and nitrofen have seriously shaken the consumer's confidence in governments' ability to ensure a proper food safety framework. Thus, the pressure on policymakers to take some significant measures towards greater food safety has risen enormously. This is noticeably reflected in the present regulatory policy. The primary principle of European food safety policy is more than ever before: 'better safe than sorry'. Hence, the precautionary principle is strictly applied in the EU, particularly in the case of genetically modified (GM) crops.[2] Since 1998, when a *de facto* moratorium was established, no further authorization of a GM crop has been granted (European Directorate General (DG) of Health and Consumer Protection, 2004). Although new regulations are in force now, a strong resistance to GM crops still persists among consumers and policymakers in the EU. In Germany for example, the question of the coexistence of GM crops with conventional and organic farming has not yet been resolved. Due to the legally unclear issue of negligence liability, no farmer currently is willing to plant GM crops. On the other hand, there are other groups in the EU which are alarmed by this biased movement against GM crops. They fear a possible loss in international competitiveness in the future, if Europe misses the chance to participate in the rapid development of GM crops.

Indeed, notwithstanding the reluctant attitudes in the EU, there is a growing number of countries trading with and cultivating GM crops. In 2003, almost 7 million farmers in 18 countries have grown GM crops on 67.7 Mha. Comparing this area with the areas of 2002 and 1996 equates to an increase of 15% and a 40 times increase, respectively. Looking at the distribution by country, the figures, not surprisingly, reveal the USA as the leading country, with a proportion of the global area of 63%, followed by Argentina (21%), Canada (6%), Brazil and China (both 4%). Among GM crops, soybeans and maize are the most widespread, with the proportion of global area sown with GM crops of 61 and 23%, respectively (James, 2003). Therefore, a focus particularly on these two crops seems to be appropriate in this chapter. Moreover, maize and soybeans are important trade products, with a global trade proportion of the world production of 15 and 32%, respectively (Food and Agriculture Organization, 2004). On this point, it should be noted that changing policies and consumer attitudes in the main export and import countries are likely to have significant impact on the world markets of both products. Particularly the USA as well as other maize- and soybean-exporting countries such as Brazil are relatively highly dependent on their markets in Europe and Japan. Based on the assumption that different perceptions pertaining to the risks of GM crops will be reflected in different regulatory and social frameworks, one could consequently conclude that new technologies will be adopted at a different pace. Regarding the ongoing adoption of GM crops in other countries, the EU has different policy responses available. Depending on the chosen strategy, it could cause some severe tensions in international trade. Indeed, in the case of GM crops, the EU has already been accused by the USA, Canada and Argentina of using this issue partly as an excuse for replacing its price support policy (Nielson and Anderson, 2000; World Trade Organization, 2004). Furthermore the choice of a certain policy option could also lead to serious arguments between different pressure groups within the EU. Nevertheless, the extent of those arguments will be determined by the economic effects of the selected strategy.

The objective of this chapter is therefore to examine the economic effects of the different policy responses from the EU regarding the introduction of GM crops in other regions. For this purpose, the chapter is organized in five parts. After a short introduction, some light will be shed on the different regulatory systems in Brazil, China, the EU and the USA. This picture will be supplemented by some general comments on recent developments in international agricultural policy. To have a close look at the different national frameworks is crucial, because they set the conditions for trade, production and consumption and therefore might have significant impact on the world market. A theoretical background is developed and will be fortified with empirical data pertaining to production and trade patterns of maize and soybeans. Thereafter the multicommodity, multiregion partial equilibrium model and the calculated scenarios will be introduced. In the following section, the simulation results will be presented and in the final section the main findings will be concluded.

Regulatory Policy Framework for GM Crops

Taking a closer look at the regulatory and agricultural policy framework is very important in order to understand why a certain technology is adopted in one country and in the other it is not. Furthermore, the specific conditions in a country always determine the direction and the magnitude of the possible impact of a new technology. Therefore, a detailed report will be given on the diverse national frameworks regarding biotechnology, with some general comments on developments in international agricultural policy at the end of this section.

Regulatory system in the EU

There is an ongoing global controversy about the possible adverse effects of GM crops on human health and the environment. In this regard, the risk perceptions between countries diverge widely, which is partly reflected in more or less rigidly established regulatory frameworks. The EU's regulatory framework, for instance, is counted among those with the most rigid systems. Indeed, the EU applies the precautionary principle very strictly, mainly because of previous serious experiences with food scares, such as BSE, nitrofen and dioxin. Indeed, with the BSE crisis, consumers have lost their confidence not only in state regulatory systems but also partly in science itself. As a result, natural consumer concerns about a new technology were exacerbated in the case of genetic modifications. Consequently, the food safety regulatory system in the EU was completely overhauled (Haniotis, 2001). The precautionary principle itself is anchored in the Cartagena Protocol on Biosafety (Cartagena Protocol on Biosafety, 2000). It provides the possibility for governments to take certain risk management measures, even though scientific uncertainty pertaining to the existing but potential risk to life or health persists (Regulation no. 1830/2003). Closely connected with the precautionary principle is the EU's general principle for food safety policy. It is an integrated 'farm to fork' approach, which means a guaranteed high level of consumer health protection at all stages of the production chain (Regulation no. 178/2001). In order to achieve this objective, the European Food Safety Authority was legally established in 2002. This authority is considered to be an independent and objective source of advice on food safety issues and is therefore responsible for the risk assessment of GM crops on the EU level (European Food Safety Authority, 2004). In contrast, the responsibility for risk management lies with the EU institutions, i.e. for horizontal legislation the Directorate General XI 'Environment, Nuclear Safety and Civil Protection' and for specific product legislation other Directorate Generals (Industry, Agriculture) (Robert Koch Institute, 2004). After the BSE crisis, the developments in food safety issues resulted in October 1998 in an authorization halt regarding genetically modified organisms (GMOs), which has become well known under the term of '*a de facto moratorium*'. Nevertheless, before the establishment of the moratorium, 18 GMOs had already been authorized under the directive no. 90/220/EEC and among them one can find a glyphosate-resistant (GR) soybean from Monsanto for the purpose of import and commercialization (European Directorate General (DG) of Health and Consumer Protection, 2004). However, in October 2002, the directive no. 90/220/EEC was replaced by the new directive no. 2001/18/EC. This regulates the deliberate release into the environment and the placing on markets of GMOs.[3] The directive's main objective is to protect human health and the environment. This should be ensured through specific environmental risk assessment on a case-by-case and step-by-step basis. Additional aspects under this directive comprise informing the public about every field release, the limitation of authorizations granted for 10 years and the obligatory monitoring of any possible adverse effects. On 7 November 2003 the European food safety framework was supplemented with two new regulations which have had to be applied in the EU member states since 18 April 2004. The central purpose of regulation no. 1829/2003[4] is the protection of human life and health, animal health and welfare, environment and consumer interests, in relation to GM food and feed. Therefore, GM food and feed have to be labelled properly. In those cases where the proportion is not higher than the threshold of 0.9% and the presence is adventitious or technically unavoidable, labelling of food and feed containing material which contains, consists of or is produced from GMOs is not required.[5] Furthermore, notification procedures according to the principle of *substantial equivalence* are no longer allowed. That means that although a GMO might be substantially equivalent to existing food or food ingredients, a simplified procedure is not permitted any more.[6] The second new regulation no. 1830/2003 seeks for the provision of a framework for the traceability of products consisting of or containing GMOs, as well as food and feed produced from them. Its objective is to facilitate accurate labelling, monitor the effects on the environment and health, and

implement appropriate risk management measures in accordance with the precautionary principle. Despite these new regulations, there are still many unanswered questions especially regarding their interpretation. A very delicate issue is the question of the coexistence of GM crops with conventional and organic farming, which has to be implemented by every member state itself. In Germany, for example, no farmer currently is willing to plant GM crops, due to the legally unclear issue of negligence liability. However, with the approval of Syngenta's Bt maize for importation and commercialization purposes, the EU lifts its 6 year *de facto* moratorium and it remains to be seen whether the case of Bt-11 maize will be proved to be an icebreaker in the EU's approval procedure (Transgen, 2004b). A lot will also strongly depend on consumers' reactions pertaining to the new developments regarding GMOs.

Regulatory system in the USA

In contrast to the EU, the USA has applied a product-oriented assessment approach for GMOs. Its regulatory system is based on the assumption that biotechnological techniques are not inherently risky and that therefore the issue of GMOs should not be regulated as a process. The assessment should rather be based on a product-oriented case-by-case approach. In the USA, three federal agencies are in charge of the regulatory oversight of agricultural biotechnology: the United States Department of Agriculture (USDA) with its Animal and Plant Health Inspection Service (APHIS), the US Food and Drug Administration (FDA) and the Environmental Protection Agency (EPA) (United States Department of Agriculture, 2004a). The objective of APHIS is to protect the agricultural environment against pests and diseases. Under the authority of the Federal Plant Pest Act, it therefore regulates the introduction, i.e. importation, movement and the environmental release, of organisms and products (regulated articles) that are known or suspected to be plant pests or pose such a risk. The EPA has the responsibility to regulate the distribution, sale and use of pesticidal substances. Besides chemically produced substances, this includes those pesticides which are produced by GMOs (so called plant-incorporated protectants, such as Bt maize). EPAs' risk assessment comprises extensive examinations regarding the potential risks to human health and the environment (gene flow, non-target organisms). The third agency, the FDA, has to make sure that all food and feed derived from plants, whether they are developed through genetic engineering or not, are safe and properly labelled. In the USA, labelling is not required as long as the GM product does not pose any health or environmental risk, for example the presence of an allergen (United States Department of Agriculture, 2004a). The basic regulatory policy of the FDA is the concept of *substantial equivalence*. GM products, which have no substantially different characteristics in comparison with their conventional counterparts, are treated like their counterparts as in the case of labelling (Venturini, 2003). If the deliberately added substances lead to products with significant differences in structure and function, only then will they have to be declared as food additives (United States Department of Agriculture, 2004a). Contrary to the very reluctant developments in the EU, in the USA the number of approvals is growing and accounts by now for 60 GMOs, among them GM maize, soybeans, sugarbeet and cotton (AgBiosDatabase, 2004). However, not only is the number of approvals rising, but the percentage of farmers planting GM crops is also increasing.

Agriculture (MOA) and the Ministry of Public Health (MPH). Whereas the MOST is responsible for the general management of biosafety, the MOA accounts for the formulation and implementation of biosafety. In 1996, the MOA issued the Implementation Regulation on Agricultural Genetic Engineering (Huang *et al.*, 2001). Under the authority of the 'Food Health Law of the People's Republic of China', the MPH is in charge of monitoring and managing food safety issues. GM food belongs to the category of 'novel food' and must be approved by the MPH (Huang *et al.*, 2002). Since 1997, four crops have been approved for commercial release: Bt cotton, tomatoes, sweet pepper and petunia, of which Bt cotton is the most important. Although other crops including rice, maize and wheat are already in the research pipeline, Chinese policymakers have not been unaffected by the international controversy regarding GM crops (Marchant *et al.*, 2002). Hence, in May 2001, the State Council issued new guidelines for Safety Administration of Agricultural GMOs, which have been supplemented by three detailed regulations on: (i) biosafety management; (ii) trade; and (iii) labelling of GM products. As a consequence, importers of GM products are now required to apply for a safety verification approval from the MOA (Huang *et al.*, 2002; Huang and Wang, 2003).

Regulatory system in China

The first worldwide commercial release of a GM crop was in China with a transgenic tobacco in 1992 (Huang *et al.*, 2002). This is not surprising, since agricultural biotechnology has been strongly supported in China since the mid-1980s. The main objective of this strong promotion was the improvement of its food security through higher yields and an improved quality of crop plants (Huang *et al.*, 2001; Marchant *et al.*, 2002). At the end of 1993, the first general regulation on biosafety was passed by the Ministry of Science and Technology (MOST). This regulation was indeed very general and therefore consisted only of some general guidelines for the responsible ministries about safety classes, application, approval and safety control measures. There are chiefly three ministries in charge of biosafety management in China: MOST, the Ministry of

Regulatory system in Brazil

Export-dependent countries such as Brazil are always looking carefully at the developments in the marketplaces of their produce, with the EU being one of the most important ones. In the case of soybeans especially, Brazil tried to uphold its claim of being a GMO-free zone for a long time. The strategy behind this was to respond properly to the critical consumer perception regarding GMOs in Europe, thus gaining a market share in world soybean exports. However, as will be reported later, at the end of 2003 this dam had finally been breached. In 1995, the National Technical Biosafety Commission (CTNBio) was established. This is Brazil's national authority pertaining to risk assessment of GMOs as well as of food and feed derived from these. It also has the responsibility for laboratory and field experiments. In 1998, the commission approved Monsanto's

'Roundup Ready' soybean as 'not harmful to the environment' (de Albuquerque Possas, 2002). However, due to strong resistance and criticism from environmental groups and consumer organizations, a court prohibited the commercial use of GM soybeans based on the precautionary principle. For reasons of judicial uncertainty, the following period from 1998 until June 2003 is sometimes referred to as the time of an involuntary moratorium. However, the true status of GM soybeans in Brazil has changed significantly. As a consequence of the judicial impasse, transgenic seed was smuggled in from Argentina, resulting in a tremendous increase in the proportion of GM soybeans by 70% in one of Brazil's biggest production areas (Cardoso, 2003). In order to solve the serious problem of illegal planting, the new government decided to legalize the sale of those illegally planted GM crops in June 2003. This decision was supplemented by a temporary measure in September 2003, whose validity initially was limited up until the following year and later extended to 2005. Besides the sale of GM soybeans, the temporary measure also allows their planting, presuming that farmers enrol themselves on lists and assume the liability for any adverse effects on the environment. Despite these developments towards a more GM-friendly environment, the federal state of Paraná in the north of Brazil is strongly opposed to GM crops. Thus it has banned the planting and transportation of GM soybeans to its state and declared itself a GM-free zone. Although the courts have stated that this procedure is illegitimate, the federal state insists on making its harbour unavailable for GM exports. In a later settlement, it was agreed to accept crops and food with a threshold level of 0.9% of GMOs. In the rest of Brazil, mandatory labelling is required for food containing more than 1% of GM material (Transgen, 2004a).

Finally, one could summarize, that the USA is the most open country regarding GM innovations, followed by China and Brazil. The EU marks the tail-light of this group of countries, with the most rigid regulatory framework for GM crops.

After the detailed remarks about the differences in the regulatory systems for GM crops, the illustration of the country-specific settings should be completed by some general comments on recent developments in international agricultural policy. With the World Trade Organization, member countries have significantly increased their pressure on countries which still have highly supported markets. Thus, it has become increasingly difficult for countries to justify a highly supported agriculture. Considerable changes in agricultural policy have already taken place in the EU, China and the USA. This process is likely to be continued, in order to promote the further liberalization of the markets.

Theoretical and Empirical Background

Technological change is a change in a production process which results from the application of scientific and applied knowledge. If technological change occurs in a certain production, productivities are changing. Following the neoclassical theory of production, productivity is determined by three factors: the state of technology; the quantities and types of resources used as inputs (the scale of production); and the efficiency with which those inputs are used (Antle and McGuckin, 1993). Surely, the three determinants of a productivity change are likely to affect each other.

Since the simulation model used only deals with demand and supply functions,[7] with the help of an illustration, the relationship between production and supply function should be emphasized at this point. Figure 14.1 shows three different diagrams to point out the connections between the production function, the cost function and the supply function. Starting with the middle diagram B, one can see a neoclassical production function (P^0), which represents the geometrical place of all technically efficient combinations of inputs and outputs by a given state of technology. Before technological change, thus, input x^0 is used to produce an output of q^0. Following the vertical solid line up to the diagram A, one can easily read the appropriate amount of costs (c^0). In addition, a linear revenue curve (R) is illustrated in A and, thus, the optimal production quantity can be found graphically. This is at q^0, where the distance between the revenue curve and the cost curve C^0 is the greatest. The third diagram C shows the appropriate marginal costs curve (MC^0) and the marginal revenue curve (MR), which intersect each other at q^0. If technological change occurs in terms of increased yields, the production function P^0 shifts to P^1. This also leads to a shift of the cost function from C^0 to C^1 and of the appropriate

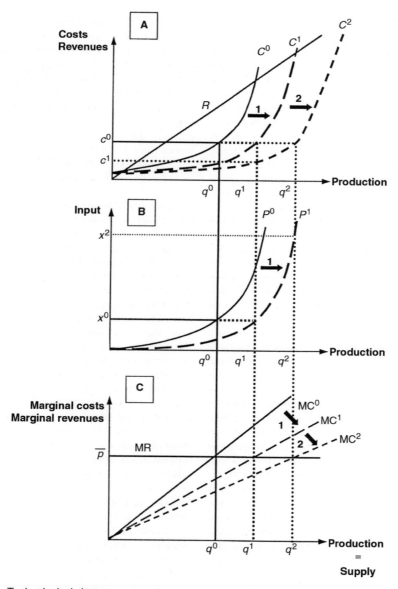

Fig. 14.1. Technological change.

marginal cost function from MC^0 to MC^1. Due to the new technology, the input productivity has increased, i.e. the higher output q^1 can now be produced with the same input x^0 at the same costs c^0 (but declining per unit costs). It is also possible that technological change only affects the input costs, as happens primarily in the case of GR soybean production. As a result, only the cost function will be shifted, but the production function remains the same. This is illustrated in diagram A via a second shift of the cost function from C^1 to C^2. With cost-reducing technological change, the same production quantity q^1 could be produced at lower cost c^1, or a greater quantity q^2 could be produced at the same costs c^0. Deriving the cost function C^2 will lead to the marginal costs MC^2 and, thus, results in an appropriate second shift of the marginal cost curve. In general, higher seed costs are also associated with new technology, as is in the case of Bt maize and GR soybeans. In our figure, this fact could be reflected in a backward shift of the cost function from C^2 to C^1.

So far, we have only considered the supply curve being affected by technological change. However, with the introduction of new technology, new improved products could also be created with a higher nutritional value, for example. If consumers perceive those new products as being more valuable and therefore are willing to pay a price premium, the demand curve will be shifted upward (Nelson and Bullock, 2001). Nevertheless, in this chapter, the focus will only be on technological change that will affect the supply curve and thus might bring consumers only indirect benefit via decreased product prices. The first generation of GM products, crops with so-called *input traits*, offer their benefits primarily to farmers. Those benefits include, for example, higher yields, reduced herbicide and insecticide use and easier weed and pest management. Implementing the *input trait* 'resistance against serious pests' into a crop has been achieved with the use of genetic engineering methods. With its help, a gene from the soil bacterium *Bacillus thuringiensis* (Bt) has been transferred successfully into the genome of different crops, such as maize and cotton. This gene enables a Bt plant to defend itself against specific insects by producing a toxic protein for this pest (Nelson, 2001). Another group of GM crops are resistant to the non-selective herbicide glyphosate, which inhibits a certain enzyme necessary for the plant's growth. While the weeds are destroyed by the herbicide treatment, the crop will survive. This is possible due to a gene transfer from *Agrobacterium* into the genome of certain crops, such as soybeans, maize and cotton, that enables the crop to express a glyphosate-resistant (GR) enzyme (Nelson, 2001).

The main benefit from using GR soybeans is a reduction in the average costs. These savings result from the relatively lower price of glyphosate in comparison with other herbicides. Furthermore, because glyphosate is a non-selective herbicide, it refers to a broad range of weed effectiveness. Hence, with the sole use of glyphosate, a mixture of other herbicides can be replaced and thus farmers reduce the time spent assessing the weed situation and make fewer trips over the field with the sprayer (Gianessi *et al.*, 2002). Additionally, due to its greater effectiveness in comparison with other herbicides, glyphosate gives farmers greater time flexibility to carry out the herbicide treatment (Carpenter and Gianessi, 2001). This makes weed management easier for the farmer and saves him managerial costs (Bullock and Nitsi, 2001). Bullock and Nitsi (2001) estimated an average cost saving for using GR soybeans of US$2.02/ha, although managerial cost savings were not included. This indicates that the actual cost savings might be higher than US$2.02/ha. Indeed, Marra *et al.* (1998), Rosegrant (2001) and Moschini *et al.* (2000) estimated a higher cost reduction for using GR soybeans instead of conventional soybeans, of US$2.43, US$3.24 and US$3.28/ha, respectively. Nevertheless, it should be noted that the actual cost savings farmers will achieve will depend on the particular weed situation and on other factors, such as the farm size and the cost structure of the farm. An interesting but unsurprising aspect is that due to the introduction of GR soybeans, those farmers growing conventional soybeans have also benefited through decreased prices of conventional herbicides (Bullock and Nitsi, 2001). Besides the potential gains from reduced herbicide costs, farmers are also faced with higher seed costs and possible yield effects. In the USA, seed for GR soybeans is up to 40% more expensive than that of conventional soybeans (Fernandez-Cornejo and McBride, 2000; Moschini *et al.*, 2000). Regarding the impact on yields, the results are ambiguous. Whereas some studies such as Carpenter (2001) indicate that the yields of GR soybeans are slightly less than the ones from conventional varieties, others found a very small increase in GR soybean yields (Fernandez-Cornejo and McBride, 2000).

While in the case of GR soybeans farmers primarily benefit from lower herbicide and managerial costs, farmers planting Bt maize do not profit from such an input cost reduction in comparison with non-adopters. This can be explained by the relatively sophisticated and costly pesticide treatment of non-Bt maize, which makes it economically undesirable for farmers to spray non-Bt maize for European corn borer (ECB) (Bullock and Nitsi, 2001). Indeed, in the USA, conventional maize fields were barely sprayed for ECB. Therefore, the impact of Bt maize is not primarily the reduced pesticide use, but rather the higher yields, as a result of actual control of ECB (Carpenter and Gianessi, 2001). Bullock and Nitsi (2001) estimated an average yield loss of conventional maize, due to the ECB, of 3.53%, which equals a decrease of 4.84 kg/ha. Yield losses due to ECB could be significant, but depend greatly on the level of infestation, which in turn is affected greatly by the weather conditions. However, interpreting the yield decrease of conventional maize as the potential yield gains from using Bt maize and assuming an average maize price of US$0.08/kg (US$2/bu), the estimated cost savings using Bt maize could be about US$3.81/ha.[8] This figure is consistent with the findings of Marra *et al.* (1998), who estimated the cost reduction applying to Bt maize at US$3.85/ha. However, Alston *et al.* (2003) estimated a benefit per hectare of US$3.27–US$11.91, subject to the level of ECB infestation. Thus, with varying planting conditions, possible cost savings that could be realized by farmers adopting Bt maize technology will also vary enormously in terms of geography and time (Bullock and Nitsi, 2001). Nevertheless, in order to estimate the possible economic impact of GM crops, we will have to make some simplifying assumptions about the cost savings. This will be done below when the scenarios will be formulated.

Finally, it can be concluded that farmers' profits could rise with both GR soybeans and Bt maize. In the case of GR soybeans, this will be due to reduced input costs, and for Bt maize as a result of increased yields. Regardless of what the source of increased profit might be, the potential of higher profits alone is a strong motivation behind farmers' decisions to adopt or not adopt new technology (Gaisford *et al.*, 2001). Besides the farm-level effects described above, a widespread adoption of new technology will also have significant market and non-market effects (Nelson, 2001). Therefore, with the introduction of new technology in a certain product process, prices and quantities of competitive and/or complementary goods might be affected, i.e. it must be taken into consideration that markets are inter-related. Thus, if the price of one product is reduced, the production of other products might become relatively more attractive. New technology will also have an impact on the upstream and downstream markets, as the experience with GR soybeans has shown. Due to the introduction of GR soybeans, prices for herbicides other than glyphosate decreased and, thus, non-adopting farmers have also gained from the new technology. If markets in one country are affected by the introduction of a new technology

and this country is an important global player on world markets, changes will not only be limited within the country. Other countries might be significantly affected, depending on their degree of international linkage.

These theoretical remarks will now be supplemented by sketching a picture of the recent developments on the world markets for soybeans and maize. In the year 2002/03, world net trade for soybeans accounted for approximately 53.8 Mt. The leading export nations were the USA and Brazil followed by Argentina, with a proportion of the global soybean trade of 45, 37 and 17%, respectively. The EU and China were the biggest net importers of soybeans with 35 and 26% of global net imports, respectively (Food and Agricultural Policy Research Institute, 2003). Taking a closer look at the origin of the European imports, the figures reveal Brazil with 51% and the USA with 39% as the main sources of soybeans for the EU (United States Department of Agriculture, 2002). As a consequence of an import restriction for soybeans by the EU, it should therefore be expected that both of these countries would be hit hardest. Adoption rates in both countries are not 100%, indicating that not all exports from these regions are GM soybeans. Nevertheless, adoption rates are rising and, without a reliable segregation system, it is more likely that those imports will be banned rather than allowed to be imported by the EU.[9] As China is itself a net importer of soybeans, it is assumed that an EU import restriction will only have a significant economic impact on China through changes in world market prices.

With maize, the picture changes noticeably: total net exports in 2002/03 accounted for roughly 69.5 Mt, whereas the USA was by far the biggest net exporter with a proportion of 67% of global net exports. There were only two other countries with noteworthy net exports of maize: China with 10.9 Mt (15.7%) and Argentina with 8.6 Mt (12.5%) (Food and Agricultural Policy Research Institute, 2004). The largest proportion of Chinese exports goes to South Korea (nearly 50%) and no significant amounts are exported to the EU (United States Department of Agriculture, 2004b). As well as China's exports, only insignificant amounts of maize are exported from the USA to the EU. They are destined primarily for Japan (32%), Mexico, Taiwan, Egypt and Canada (together 36%) (National Agricultural Statistic Service, 2003).

Thus, an import ban on maize is supposed to have only indirect effects via changes in world market prices for maize. Besides China and the USA, Brazil was also a net exporter of maize, but only with a relatively small fraction (1.6% or in absolute terms 1.1 Mt) of global net exports (Food and Agricultural Policy Research Institute, 2003). As 12% of Brazil's maize was exported to the EU (United States Department of Agriculture, 2004c), Brazil will not only be affected by changing world market prices, but also directly by the import restriction. As the EU accounts for net imports of maize of 2.4 Mt (3.5%), it was only a small importer in 2002/03. More important net importers of maize were Japan, South Korea and Mexico, with a proportion of global net imports of 22.3, 12.2 and 9.3%, respectively (Food and Agricultural Policy Research Institute, 2003).

Simulations and Results

In this chapter, the multicommodity, multiregion trade model AGRISIM is used to analyse the economic effects on prices, production, trade and welfare, due to different EU responses to the adoption of GM crops in other countries. The actual version of this comparative–static, partial equilibrium model is based on the 'Static World Policy Simulation Model' (SWOPSIM) from the USDA. AGRISIM covers 16 regions and nine commodities.[10] The database was recently updated for the year 2001. The simulations will only be focused on Brazil, China, the EU 15 and the USA as they are important global players in agricultural trade. Hence, any political decision in these countries could be of significant importance to the world market. As mentioned previously, it is primarily the effects on the markets of maize and soybeans which will be examined. Unfortunately, in the current version of the model, maize and soybeans are only included in commodity aggregates, i.e.

coarse grains for maize and oilseeds for soybeans. Regardless of that fact, taking a closer look at the database reveals an appropriate reflection in their aggregates for both products for Brazil, China and the USA and a medium one for the EU. Thus, one important task for further model enhancements has to be the separation of both commodities from their aggregates, especially in the case of the EU.

Three different scenarios have been formulated in order to calculate the economic effects, due to an introduction of GM crop technology in Brazil, China and the USA, with different responses from the EU. Table 14.1 gives a brief overview of the considered scenarios. In the base scenario, everything will be unchanged in order to provide a reference for the comparison with the other three scenarios. All three of the following scenarios are built up of two parts, whereas the first part always remains the same. It implies a shift in the supply curve for maize (+1.5%) and soybeans (+2.3%) due to the introduction of new genetically modified varieties in Brazil, China and the USA.

These figures have been calculated based on data from the USDA by assuming a Cobb–Douglas production function of the form:

$$q = A \cdot I_1^{\alpha 1} \cdot I_2^{\alpha 2} \cdot I_3^{\alpha 3} \cdot I_4^{\alpha 4} \cdot I_5^{\alpha 5} \cdot F^{1 - \Sigma \alpha i} \qquad (1)$$

where q = production; A = technology parameter; I_1 = seed; I_2 = fertilizer; I_3 = plant protection products; I_4 = capital; I_5 = labour; F = fixed input; and α_i = production elasticities, as well as profit maximizing supply responses with respect to input price changes and changes in total factor productivity. In that case, one can easily derive the total shift of the supply function as the weighted sum of individual shift factors:

$$\frac{dq}{q} = \frac{1}{\beta} \cdot \frac{dA}{A} - \sum \frac{\alpha_i}{\beta} \cdot \frac{dp_i}{p_i} \qquad (2)$$

where $\beta = 1 - \Sigma \alpha_i$; p_i = input prices; $i = 1, 2, 3, 4, 5$; and d = absolute change.

Table 14.1. Calculated scenarios.

Base Scenario	no supply shift	+	no response from the EU
Scenario 1	} supply shift in Brazil, China and the USA for maize and soybeans		import ban of EU for both products
Scenario 2		+	no response from the EU
Scenario 3			EU also adopts technology (supply shift)

Source: Own illustration.

As is still the case for the majority of GM crops, the EU has not yet allowed their importation nor their cultivation within the EU. This situation is reflected in scenario 1, assuming a total import ban of all maize and soybeans from Brazil, China and the USA. A total import ban was assumed to allow for the very hostile attitudes of a great deal of European consumers towards GMOs and the near non-existence of a reliable segregation system for GM and non-GM crops in these three countries. Because this model is not able to distinguish where the imports of one country are coming from, the net protection rate (NPR) has served as a proxy variable, i.e. the NPR has been increased in order to simulate an import ban. In the second scenario, the EU lifts the import bans, but does not approve any GM crop for cultivation purposes. In the last scenario, besides Brazil, China and the USA, the EU also fully adopts Bt maize and GR soybean technology and thus also experiences a shift in the supply curve.

Tables 14.2–14.4 report the simulation results for Brazil, China, the EU and the USA regarding production, trade and price effects for coarse grains and oilseeds and the overall welfare effects. First, some attention should be given to the overall welfare effects due to the different EU responses (see Table 14.2). Starting with scenario 2, the pure effect of a technology-induced supply shift in the non-EU countries could be easily seen. It leads to increases in total welfare for all countries including the EU but, in particular, producers from the adopting countries would gain a significant amount of between US$91.45 million and US$305.71 million. If the EU also adopts the new technology (see scenario 3), the primary winners will be European producers (US$198.9 million), signifying an improved international competitiveness. On the other hand, due to negative effects on the EU's budget, the total welfare effects for the EU in this scenario are slightly less than those in scenario 2. Comparing these results with scenario 1 reveals high losses for all producers (US$34.5–US$349.0 million) except for those from the EU (US$2606.7 million). This is the result of the artificially high protection of European producers, which in turn also affects other producers via depressed world market prices. Nevertheless, the gains for consumers outweigh the losses of producers, and thus total welfare becomes positive for Brazil, China and the USA. Thereby China's enormous welfare improvement is a consequence of the huge benefit for the consumers.

Table 14.2. Welfare effects in mill. US$

		Scenario 1	Scenario 2	Scenario 3
Brazil	Producer	−55.77	91.45	85.31
	Consumer	96.97	13.23	15.48
	Budget	10.32	−7.00	−6.36
	Total	51.53	97.68	94.42
China	Producer	−349.02	209.37	201.06
	Consumer	679.84	57.49	68.88
	Budget	108.87	−22.12	−21.40
	Total	439.69	244.74	248.54
EU	Producer	2606.74	−8.28	198.90
	Consumer	−2931.65	17.75	20.87
	Budget	−435.01	25.87	−188.26
	Total	−759.92	35.34	31.51
USA	Producer	−34.14	305.71	280.00
	Consumer	290.55	61.69	74.29
	Budget	−126.78	−171.07	−161.42
	Total	129.63	196.33	192.87

Source: Own calculations.

Table 14.3. Simulation results for coarse grains in %.

		Scenario 1	Scenario 2	Scenario 3
Production	BRA	1.17	1.17	1.08
	CHI	1.25	1.27	1.22
	E15	−0.21	−0.36	1.04
	USA	1.02	1.08	0.97
Farm Gate Price	BRA	−1.01	−0.80	−0.99
	CHI	−1.01	−0.80	−0.99
	E15	−0.61	−0.64	−0.80
	USA	−1.01	−0.80	−0.99
Net Trade	BRA	−24.02	−26.77	−20.52
	CHI	−73.74	−68.15	−60.89
	E15	−9.97	−7.18	10.28
	USA	4.54	4.86	4.22

Source: Own calculations.

Table 14.4. Simulation results for oilseeds in %.

		Scenario 1	Scenario 2	Scenario 3
Production	BRA	0.92	1.95	1.89
	CHI	2.05	2.23	2.22
	E15	21.64	−0.24	2.02
	USA	1.03	1.98	1.92
Farm Gate Price	BRA	−2.39	−0.64	−0.75
	CHI	−2.39	−0.64	−0.75
	E15	78.96	−0.73	−0.86
	USA	−2.39	−0.64	−0.75
Net Trade	BRA	0.70	2.99	2.86
	CHI	−1.67	−3.34	−3.24
	E15	−78.75	1.01	−2.58
	USA	0.31	4.26	4.02

Source: Own calculations.

Now taking a closer look at the market effects for coarse grains, increases in production of between 0.97 and 1.27% for Brazil, China and the USA are shown, whereas the growth in production is highest in the second scenario (see Table 14.3). The EU's coarse grain production decreases slightly in the first two scenarios, but increases in the last scenario by about 1.04% as a result of the introduction of the new GM technology. The changes in net trade for coarse grains have been significant, especially for Brazil and China. Since both countries are net importers of coarse grains in AGRISIM's base year 2001, a fall in net trade indicates either extended exports or reduced imports, which is due to the increased production of coarse grains.

For scenario 1 and 2, the EU's net trade decreases,[11] because consumption within the Union increases as a consequence of a decline in consumption prices. Owing to the adoption of GM maize, not only does the production expand but also exports of coarse grains increase (change in net trade +10.3%).

Just as in the case of coarse grains, the technology-induced supply shift for oilseeds leads to a rise in production for all three adopting countries of between +0.92 and +2.23% in all three scenarios (see Table 14.4). Particularly in scenario

1, oilseed production in the EU responds with a significant extension of about +22%. This considerable increase is primarily the effect of the import ban, which protects domestic markets from foreign competitors and thus leads to a strong rise in farm gate prices of +79%. As a consequence of the big price increases, consumption falls and, together with the development in production, leads to a lower import demand for oilseed −78.8%.[12] Due to the reduced import demand of the EU, farm gate prices are affected and decrease (−2.39%) more in this scenario than in the second (−0.64%) or third (−0.78%) scenario for Brazil, China and the USA. As a result, despite the shift of the supply curve, the production increase is only minor (see scenario 3). As the EU lifts its import ban for oilseeds (scenario 2), farm gate prices decrease in the EU and thus production will be cut appropriately. With lower production of oilseeds in the EU, the situation on the world market eases off. This changes a little with the adoption of GM soybeans in the EU, leading, via the supply shift, to a larger supply of oilseeds, and thus yields lower farm gate prices in all four countries. The earlier assumption that Brazil and the USA in particular will be affected by an EU ban of oilseeds cannot be proven.

Conclusions

In this chapter, the economic effects on prices, production, trade and welfare were calculated by the partial equilibrium model AGRISIM. The simulations gave a first hint of the economic effects due to the introduction of a new technology while taking into account different policy responses of the EU. The results indicate a total welfare gain due to the introduction of GM crops for adopting and non-adopting countries. With the adoption of GM crop technology, in particular producers gain and among them mostly early innovators. An import ban on GM maize and soybeans by the EU will negatively affect producers in adopting countries. However, due to the relatively higher gains of consumers, the overall welfare in adopting countries is still positive with an import ban, but, except for China, significantly higher in the two scenarios without a ban.

For the improvement of the model towards the specific circumstances of GM crops, i.e. a more appropriate organization of the model's regions and commodity aggregates, further efforts are required. Moreover, one should contemplate a separation of GM and non-GM crops in the model. Additional fields are the integration of externalities, risk, uncertainty and irreversibility into the model.

Notes

[1] In this chapter, the European Union refers to the EU-15. The ten new members are not considered.

[2] In this chapter, we assume GM crops to be defined as genetically modified organisms (GMO), in which the genetic material has been altered in a way that does not occur naturally by mating and/or natural recombination (Regulation no. 1829/2003/EC).

[3] GMOs are seed and the augmentable parts of the crop.

[4] Regulations regarding GM food were separated from Regulation no. 258/97.

[5] Products derived from animals fed with GM feed do not have to be labelled, e.g. milk and milk products, beef and eggs.

[6] This procedure was permitted in the novel food Regulation no. 258/97. However, in Regulation no. 1829/2003 (6), the EU argues that the '*substantial equivalence* is a keystone in the procedure for assessment of the safety of GM food, but that it is not a safety assessment in itself '.

[7] For detailed information about the structure of the model and its equations, see Schmitz (2002).

[8] $9.4/acre are the result of taking the calculated yield increase of 4.7 bu/acre from Bullock and Nitsi (2001) times the product price of $2/bu results.

[9] Instead of an import ban, GM soybeans from Brazil and the USA could be boycotted by consumers in the EU, thus also leading to reduced imports into the EU.

[10] It comprises the countries: Australia, Brazil, China, Cuba, India, Japan, Mexico, South Africa, Thailand, Ukraine, the USA and the country aggregates: Belarus and Russia, EU-15, Central European Countries (CECs), the rest of the OECD and the rest of world. Why exactly these regions have been chosen can be explained by the original purpose of AGRISIM to analyse the impact of different policy scenarios for the world sugar market. The nine products include: wheat, coarse grain, rice, oilseeds, sugar, pork, poultry, milk and beef.

[11] In AGRISIM's base year, the EU is a net exporter for coarse grains.

[12] In AGRISIM's base year, the EU is a net importer for oilseeds.

References

AgBiosDatabase (2004) Accessed at http://www. agbios.com/dbase.php in April 2004.

Alston, J.M., Hyde, J., Marra, M.C. and Mitchell, P.D. (2003) *An Ex Ante Analysis of the Benefits from the Adoption of Corn Rootworm Resistant, Transgenic Corn Technology.* Presented at the General Meeting of the Australian Agricultural and Resource conomics Society, Fremantle, Western Australia.

Antle, J.M. and McGuckin, T. (1993) Technological innovation, agricultural productivity, and environmental quality. In: Carlson, G.A., Zilberman, D. and Miranowski, J.A. (eds) *Agricultural and Environmental Resource Economics.* Oxford University Press, Oxford, UK, pp. 175–220.

Bullock, D. and Nitsi, E.I. (2001) GMO adoption and private cost savings: GR soybeans and Bt corn. In: Nelson, G.C. (ed.) *Genetically Modified Organisms in Agriculture.* Academic Press, San Diego, California, pp. 21–38.

Cardoso, F. (2003) Genetically altered quagmire: Brazil's involuntary moratorium . . .. Checkbio. org (April 1), accessed at www.checkbiotech. org in May 2004.

Carpenter, J. (2001) *Comparing Roundup Ready and Conventional Soybean Yields 1999.* National Center for Food and Agricultural Policy, Washington, DC (www.ncfap.org).

Carpenter, J. and Gianessi, L.P. (2001) *Agricultural Biotechnology: Updated Benefit Estimates.* National Center for Food and Agricultural Policy, Washington, DC (www.ncfap.org).

Cartagena Protocol on Biosafety (2000) Convention on Biological Biodiversity, accessed at http://www.biodiv.org/biosafety/protocol.asp in April 2004.

de Albuquerque Possas, C. (2002) *The Future of Agricultural Biotechnology: Prospects for Brazil.* Executive Secretary-National Technical, Biosafety Commission-CTNBio, Agricultural Outlook, Forum 2002.

European Directorate General (DG) of Health and Consumer Protection (2004) (http://www. europa.eu.int/comm/food/food/biotechnology/ authorisation/index_en.htm) accessed on 18 May, 2004.

European Food Safety Authority (2004) Accessed at www.efsa.eu.int/about_efsa/catindex_en.html in April 2004.

Food and Agricultural Policy Research Institute (2004) Agricultural Outlook 2004 – world coarse grain and oilseeds. Accessed at www.fapri.org in May 2004.

Food and Agriculture Organization (2004) FAOSTAT, Agriculture, accessed at www.fao.org in May 2004.

Fernandez-Cornejo, J. and W. McBride (2000) With contributions from Klotz-Ingram, C., Jans, S. and Brooks, N. *Genetically Engineered Crops for Pest Management in U.S. Agriculture.* Agricultural Economics Report No. 786. United States Department of Agriculture.

Gaisford, J.D., Hobbs, J.E., Kerr, W.A., Perdikis, N. and Plunkett, M.S. (2001) *The Economics of Biotechnology.* Edward Elgar Publishing, Northampton, Massachusetts.

Gianessi, L.P., Silvers, C.S., Sankula, S. and Carpenter, J.E. (2002) *Plant Biotechnology: Current and Potential Impact for Improving Pest Management in U.S. Agriculture – An Analysis of 40 Case Studies.* NCFAP, Washington.

Haniotis, T. (2001) The economics of agricultural biotechnology: differences and similarities in the US and the EU. In: Nelson, G.C. (ed.) *Genetically Modified Organisms in Agriculture.* Academic Press, San Diego, California, pp. 171–178.

Huang, J. and Wang, Q. (2003) *Biotechnology Policy and Regulation in China.* IDS Working Paper 195, Institute of Development Studies, University of Sussex, UK.

Huang, J., Wang, Q. and Keeley, J. (2001) *Agricultural Biotechnology Policy Processes in China.* Institute of Development Studies, University of Sussex, UK.

Huang, J., Wang, Q., Keeley, J. and Falck-Zepeda, J. (2002) Agricultural biotechnology policy and impact in China. *Economic and Political Weekly* [India] 37 (27) (Review of Science Studies), 2756–2761.

James, C. (2003) *Status of Commercialized Transgenic Crops: 2003.* International Service for the Acquisition of Agri-biotech Application (ISAAA).

Marchant, M.A., Feng, C. and Song, B. (2002) Issues on adoption, import regulations, and policies for biotech commodities in China with a focus on soybeans. *AgBioForum* 5, 167–174.

Marra, M., Carlson, G. and Hubell, B. (1998) *Economic Impact of the First Crop Biotechnologies.* North Carolina State University, Department of Agricultural and Resource Economics.

Moschini, G., Lapan, H. and Sobolevsky, A. (2000) Roundup Ready soybeans and welfare effects in the soybean complex. *Agribusiness* 16, 33–55.

National Agricultural Statistic Service (2003) *USDA Agricultural Statistics 2003, Grain and Feed.* Accessed in April 2004 at http://www.usda.gov/nass/pubs/agr03/03ch1.pdf

Nelson, G.C. (2001) Traits and techniques of GMOs. In: Nelson, G.C. (ed.) *Genetically Modified Organisms in Agriculture.* Academic Press, San Diego, California, pp. 7–15.

Nelson, G.C. and Bullock D. (2001) The economics of technology adoption. In: Nelson, G.C. (ed.) *Genetically Modified Organisms in Agriculture.* Academic Press, San Diego, California, pp. 15–21.

Nielsen, C. and Anderson, K. (2000) *GMOs, Trade Policy and Welfare in Rich and Poor Countries.* CIES Policy Discussion Paper No. 0021. Center for International Economic Studies, University of Adelaide.

Robert Koch Institute (2004) Robert Koch Institute website accessed at http://www.rki.de/ GENTEC/GESETZ/ GESTZ.HTM in April 2004.

Rosegrant, M.W. (2001) Simulation of world market effects: the 2001 world market with and without Bt corn and GR soybeans. In: Nelson, G.C. (ed.) *Genetically Modified Organisms in Agriculture.* Academic Press, San Diego, California.

Schmitz, K. (2002) Simulationsmodell für die Weltagrarmärkte – Modellbeschreibung. In: Schmitz, P.M. *Nutzen-Kosten-Analyse Pflanzenschutz.* Wissenschaftsverlag Vauk Kiel KG, Kiel, pp. 117–137.

Transgen (2004a, Brasilien: Anbau von gv-Soja wird Normalität. TransGen 10.02.2004, accessed at www.transgen.de in May 2004.

Transgen (2004b), EU-Kommission: Bt11-Mais zugelassen. Das Ende des Moratoriums. TransGen 19.05.2004, accessed at www.transgen.de in May 2004.

United States Department of Agriculture (2002) Attaché Report no. E23144, FAS.

United States Department of Agriculture (2004a), Animal and Plant Health Inspection Service accessed at www.aphis.usda.gov/brs/usregs.htm on April 13.

United States Department of Agriculture (2004b) Attaché Report no. CH4005, FAS.

United States Department of Agriculture (2004c) Attaché Report no. BR4607, FAS.

Venturini, L. (2003) *The Political Economy of GMOs Regulation in the EU and the US: Towards a Convergence?* Paper presented on the 7th ICABR International Conference, Ravello, Italy.

World Trade Organization (2004) Homepage of the WTO. Accessed at http://www.wto.org/english/tratop_e/dispu_e/dispu_subjects_index_e.htm#gmos in April 2004.

15 Opposition to Genetically Modified Wheat and Global Food Security

Faycal Haggui, Peter W. B. Phillips and Richard S. Gray
Department of Agricultural Economics, University of Saskatchewan, 51 Campus Drive, Saskatoon, Saskatchewan, Canada S7N 5A8

Introduction

Even with the recent acceleration of investment in agrifood research, we have seen a slowdown in productivity gains and greater yield variability in some crops due to environmental and disease pressures. After rice, wheat is the most important food staple; it is widely produced and consumed worldwide. The rate of growth of wheat yields is decelerating at an increasing rate for developing and developed countries alike. This trend reinforces a predominant fear that we may be reaching a yield plateau in the case of wheat (Pingali and Heisey, 2001), and that future global supply may be inadequate to meet the future demands of an increasing world population. Sustained research is going to be needed, but conventional technologies are unlikely to reverse the trends.

Genetically modified (GM) wheat is one of many ways both to increase global wheat yields and to reduce their variability in the face of climate change. For example, GM wheat can be adapted for drought and altered climate. Higher wheat yields might enhance global food availability, ensuring a greater food supply, food stock and affordable, more stable world prices. However, due to the importance of wheat production and trade worldwide, there are serious concerns about whether GM varieties of this crop will see widespread adoption. On May 10, 2004, after more than 3 years of an increasingly tense debate about the appropriateness of GM Roundup Ready (RR)

wheat, Monsanto Company announced that it would not be commercializing the product. The company reported that it had suspended all further research and commercialization efforts, in all countries, effective immediately. While it has not withdrawn its applications currently with Canadian and US regulators, the company states that they will not proceed in the foreseeable future. Many groups hailed Monsanto's announcement as a victory for consumers and for farmers. While for some that may be true, that decision will have a mix of impacts in both the short and long term for researchers, producers and consumers. Ultimately, we have exchanged short-term certainty for long-term uncertainty.

The goal of this chapter is to evaluate the impact of consumer opposition (particularly European opposition) to the adoption of GM food (GM wheat) and to the realization of its benefits in terms of global food security. A number of scenarios are analysed in order to consider the range of potential effects for the European opposition on global food security. The specific objectives of the chapter are to: develop a conceptual model of consumer acceptance of GM crops; quantify the effects of GM crop resistance on adoption rates of GM crop; and evaluate the effects of GM crop resistance under projected climate change scenarios on global food security. This allows computing of an *ex ante* estimate of what we have potentially given up in the context of the Monsanto decision.

The rest of the chapter is organized into four sections. The first provides background regarding the food security problem, the need for technical change (GM technology), and intolerance to biotechnology, specifically GM wheat. In the following section, we identify critical factors in technology diffusion, ultimately allowing for the estimation of a consumer-responsive global GM wheat diffusion model. This permits the computation of a global food security (GFS) model that allows dynamic economic responses to opposition to GM food. The GFS model is calibrated using production, consumption and price data for wheat. In the next section, the GFS model is used to examine GFS under several opposition scenarios and under two climate change assumptions. The comparison between scenarios estimates the impacts of opposition to GM wheat on GFS. Finally, the last section offers a number of concluding observations.

Background

Food security has been defined in several ways. The Food and Agriculture Organization (1985) states that 'the ultimate objective of food security is to ensure all people, at all times, are in the position to produce or procure the basic food they need and that it should be an integral objective of economic and social plans'. More recently, food security is defined as a state of affairs where all people at all times have access to enough safe and nutritious food to maintain a healthy and active life (World Bank, 1986; Food and Agriculture Organization, 1996). The factors that determine the degree of food security in any region or zone are food availability, accessibility to food by all people in that region, and nutritional adequacy (Lacy and Busch, 1986; Food and Agriculture Organization, 1998).

The world has made significant progress over the last decade in decreasing the incidence of undernourishment in the developing world, most recently lowering the rate to 17% in 1997–1999 from 20% in 1990–1992 (Table 15.1). The total number of undernourished people in developing countries decreased by 4.8% between 1990/92 and 1997/99, equal to an average decline of 6 million per year. To achieve the 1996 World Food Summit goal to halve the number of hungry people by 2015, the annual rate of decrease of undernourished people should be 24 million, or over four times the current pace. Over 95% of undernourished people are in developing countries, while 3.6% are in countries that are in transition and 1.3% are in industrialized countries. Throughout the 1990/99 period, sub-Saharan Africa was the region with the highest incidence of undernourishment, with over one-third of the total population being undernourished. South Asia and East Asia also have high rates of undernourishment (26 and 16%, respectively). South Asia had the highest absolute number of undernourished people (39% of all undernourished people in the world), followed by sub-Saharan Africa with 25% (Table 15.1).

By 2020, wheat demand worldwide is expected to increase by 40% from its level in 1997 (Rosegrant et al., 1997), and less developed countries will double their grain imports, including wheat, by 2020 (Pinstrup-Anderson et al., 1999).

Table 15.1. Incidence of undernourishment in developing countries.

	Million people		% of total population	
	1990/92	1997/99	1990/92	1997/99
All developing countries	815	776	20%	17%
Sub-Saharan Africa	168	194	35%	34%
Near East and North Africa	25	32	8%	9%
Latin America and Caribbean	59	54	13%	11%
South Asia	289	303	26%	24%
East Asia	275	193	16%	11%

Source: Food and Agriculture Organization (2002).

A key problem facing policymakers is that yields of all major food crops have been growing less rapidly than in the past and, for some crops, yields have stagnated. Worldwide, yield growth decreased by 15% between 1961–1974 and 1975–1988, while it decreased by 64% between 1975–1988 and 1989–2002 (see Table 15.2). The same trend is observed for the European Union (EU), transition markets and developing Asia. In contrast, Central America, North America and South Africa experienced an increase in yield growth rate between the later two periods. The rest of the world (including developing Africa) experienced significant declines in yield growth between 1975–1988 and 1989–2002.

As illustrated in Table 15.3, wheat yield growth rates are decreasing at an increasing rate for the world, including developing and developed countries, as well as in most geographical regions. A simple regression of yield on time and time squared allows us to identify trends and convexities in the trends (i.e. ln(yield) = a + bt + c[1/2

Table 15.2. Annual growth rates and annual average of wheat yield, 1961–2002.

Regions	Annual average yields (hg/kg)			Annual growth rate of yields		
	1961–1974	1975–1988	1989–2002	1961–1974	1975–1988	1989–2002
World	13,945	19,980	25,943	3.29**	2.81**	1.01**
Developed countries	15,980	21,403	26,509	3.49**	2.21**	0.66*
Developing countries	10,616	17,990	25,272	3.40**	3.95**	1.49**

**Significant at 1%; *significant at 5%.
Source: authors' calculation from Food and Agriculture Organization data.

Table 15.3. Decelerating rate of wheat yield growth, 1961–2002.

Regions	Coefficients of the regressions		
	a	b	c
World	9.2752**	0.0366**	−0.0007**
Developed countries	9.4257**	0.0338**	−0.0007**
Canada and USA	9.6181**	0.0237**	−0.0006**
Developed Asia	9.9142**	0.0207**	−0.0005
EU (15)	9.8590**	0.0430**	−0.0007**
Oceania	9.3724**	0.0033	0.0003
Transition markets	9.1811**	0.0417**	−0.0012**
Developing countries	8.9336**	0.0481**	−0.0008**
Central America	9.8263**	0.0468**	−0.0012**
South America	9.3997**	0.0030	0.0007**
Developed Africa	8.6417**	0.0251**	0.0005
Developing Africa	8.9855**	0.0169**	0.0003
Developing Asia	8.8461**	0.0566**	−0.0011**

**Significant at 1%. The coefficients (a, b and c) are taken directly from regression ln(yield) = a + bt + c(1/2 t^2) for each region.
Source: authors' calculation from Food and Agriculture Organization data.

t^2]). The coefficient c is a measure of the speed of change in yield growth with time. If this coefficient is positive, yield growth is increasing with time, and vice versa. Despite the continued development of newer generations of high yielding varieties (Sayre, 1996), growth in yields has slowed (i.e. the coefficient c is negative). A variety of explanations are offered for this, including: resource degradation and pressure on the environment (Cassman and Pingali, 1995; Huang and Rozelle, 1995); lack of use of modern inputs and diminishing returns due to continuous cropping (World Bank, 1992); and exhaustion of the benefits of transition to high yielding wheat varieties increased significantly, reaching 82% in 1997 (Byerlee and Moya, 1993; CIMMYT, 1996).

At the same time as yield growth has been slowing, yields have become more volatile in some regions. Table 15.4 shows that the yield variability for wheat has decreased globally from 1961 to 2002, both for developed and developing countries.

Globally the average percentage deviation has decreased significantly from 4.1 to 2.6%. Although yield variability declined for developing countries between the 1961–1974 and 1989–2002 periods, there was a significant rise in variability in the 1975–1988 period. Other regions, such as Central America, and even North America, have observed an increase in yield variability between the 1975–1988 and 1989–2002 periods. Although this increase was not significant, it is somewhat alarming as North America and Argentina are among the biggest exporters in the world. Many former Soviet Union states are also traditional large wheat producers. A small increase in yield variability in such key regions may lead to a significant decline in food availability and a spike in wheat prices, worsening already fragile global food security.

Due to climate change and the possibility of reaching a yield plateau with conventional varieties, it is possible that the pattern of increased variability for wheat will be more common worldwide and

Table 15.4. Yield variability coefficients, 1961–2002.

Regions	Average percentage deviation from trend[a]			Tests of significance between periods[b]		
	Period 1 1961–1974	Period 2 1975–1988	Period 3 1989–2002	1 and 2 F1,2	1 and 3 F1,3	2 and 3 F2,3
World	4.1	4.1	2.6	1.34	8.54**	6.40**
Developed countries	5.6	5.1	3.9	2.22	11.03**	4.96**
Canada	20.8	14.8	11.6	2.40	4.42**	1.84
Developed Asia	22.1	7.1	8.1	2.97*	7.23**	2.43
EU (15)	5.4	6.1	4.0	0.99	6.89**	6.94**
Oceania	17.7	23.6	22.0	0.54	0.65	1.21
USA	6.9	6.1	7.7	1.62	1.79	1.11
Transition markets	11.7	11.0	12.0	2.72*	3.83*	1.41
Developing countries	3.5	5.0	2.8	0.73	4.63**	6.38**
Central America	9.6	5.6	6.2	5.49**	4.19**	0.76
South America	11.6	9.8	6.8	1.60	1.41	3.48*
Africa developed	16.5	19.7	13.7	0.95	0.54	0.57
Africa developing	10.4	10.5	8.2	0.43	1.71	4.01**
Asia developing	4.2	5.0	2.9	1.08	6.32**	5.86**

[a]Standard error from regression: ln(yield) = a + bt.
[b]Significance is tested using F-test two-sample for variances.
**Significant at 1%; *significant at 5%.
Source: authors' calculation from Food and Agriculture Organization data.

more significant in the future. Climate change may already be having an effect. In 2002/03, Australia and Canada were devastated by drought, while East Europe (especially Romania and Hungary) were hard hit by a heat wave. Adverse weather also affected US wheat in 2002/03. Some researchers link these events to climate change. Adverse weather may reduce not only wheat yields but also wheat acreage because bad weather could discourage farmers from continuing to farm.

The decrease of global acreage, decelerating yield growth and increase of yield variance in several wheat-producing countries suggest that future wheat supplies may be less stable and may not meet the demand of the increasing global population. Pinstrup-Andersen and Schioler (2001) argue that conventional technologies might contribute about 70% of the technological advances needed to meet the world food programme targets. Biotechnology offers one way to make up the shortfall. Better understanding of the wheat gene map, use of molecular markers to accelerate the development of better wheat varieties and transgenic modifications to incorporate new traits could help to reverse or slow these worrying trends. Global GM wheat research started around 1993 and its trends have been closely related to the degree of acceptance of GM food technology. Field trial data from the USA and Canada reveal that a variety of traits are being worked on, including herbicide tolerance (i.e. RR tolerance), *Fusarium* resistance, viral resistance, abiotic stress resistance (e.g. drought, salinity, heat and cold) and various product quality improvements (e.g. improved digestibility, increased yield, altered storage proteins and lower glutens).

The future of GM wheat appears to be inextricably linked to market responses in advanced industrial economies. EU opposition to most GM agrifood products appears to have affected the outlook for GM food through several ways. The EU demands that food exporters to the EU should label their products, which often raises either costs or risks for the trade. There is some evidence that the requirement for labelling in the EU has already affected Pakistan (*Daily Times*, 25 August, 2003). There are reports that the EU threatened to stop importing agricultural products from Pakistan and other countries if these countries did not implement labelling rules on feed and food exported to the EU. This opposition to GM food has had some unexpected impacts. In summer 2002, Zambia and Zimbabwe both rejected food aid despite widespread famine in those countries. The Zambian decision to reject food aid for 3 million people was based on concerns over health and economic implications of GM food (BBC News, 17 August 2002). The Zimbabwean government has also rejected food aid of American maize due to concerns over the presence of GM seeds, which if planted could 'contaminate' their crops and jeopardize the country's exports to countries that accept only GM-free foods (Weiss, 2002). In another example from Africa, Uganda refused to grow a disease-resistant type of banana, fearing it would cost them exports to Europe (Angelo *et al.*, 2003). India also rejected a large shipment of maize–soy blended food aid from the USA in January 2003 because it contained GM food (Luce, 2003).

When GM wheat entered the regulatory process in the USA and Canada, many importers indicated unease with GM wheat in the trade. The Canadian Wheat Board (2003) reported that 82% of its customers indicated that they will reject the import of GM wheat, while US Wheat Associates (2002) reported that an Asian wheat market survey revealed that all of the Japanese, Chinese and Korean, 82% of Taiwanese and 78% of South Asian buyers had indicated that they will reject GM wheat. Most buyers also indicated they had zero tolerance to even trace contamination, regardless of government regulations. In contrast, a few developing Asian markets, such as Indonesia, indicated they would accept GM wheat, since they expect it to be cheaper than the conventional varieties (Cropchoice, 2003). Meanwhile, a Eurobarometer survey by the European Commission in 2003 found that more than 68% of consumers in countries that joined the EU in 2004 reject GM food, with more than half believing GM food to be dangerous (*The Economist*, 2003).

On 10 May 2004, Monsanto announced that it would cease to develop and commercialize its RR herbicide-tolerant variety of GM wheat. In effect, the technology has been shelved for a number of years, if not permanently. While other firms have not yet followed suit, there is substantial risk that other biotechnology research may slow or stop until market prospects improve. The remainder of this chapter undertakes a modelling exercise – modifying the adoption decision to allow for direct feedbacks from consumer markets – to illustrate the array of potential outcomes.

Theory and Model

This section deals with modelling of the impact of GM market acceptance on GM adoption rates and, ultimately, GFS. A production model is used to estimate the impact of GM market acceptance (or lack thereof) on GM adoption rates and the impact this will have on supply. A stochastic simulation model is then used to estimate the impacts on price levels and future global food security. By construction, the model ignores many exogenous variables that will impact food security but are not affected by the

GM acceptance. Figure 15.1 outlines the nature of the model.

A market model is combined with a GM adoption model to create a GFS model, which allows dynamic economic responses to food production shocks, such as climate change. In the adoption model, the rate of adoption of GM wheat is affected by major importers' policies through trade. The adoption rate depends on the diffusion of this technology, which is driven by consumer perception of the technology. The market model consists of both a supply and a demand side. Any year's wheat supply is the sum of the previous

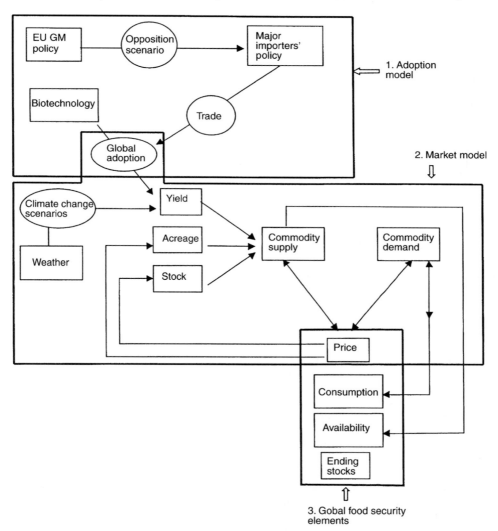

Fig. 15.1.   Global food security model.

year's stock and the current year's production. The quantity of wheat stock each year is estimated using a stock equation. Production is divided into a yield forecast and acreage responses. The acreage function permits the estimation of the total acreage planted with wheat. The yield forecast depends on assumptions about the productivity increases for GM and conventional wheat and the adoption rate of GM wheat. Yield forecasts of conventional wheat are estimated from the historical data. The yield forecast of GM wheat is estimated under specific assumptions. The realized yields depend on the climate and environmental shocks assumed to be random in this study. The demand side consists of consumption determined by a linear equation.

The adoption and market model are combined to compute the food security elements (Fig. 15.1). The level of food security is assessed using several factors, including: real wheat prices, which are a good measure of food security as lower real wheat prices translate into cheaper imports for developing countries; consumption, which measures whether access to food has improved or not; ending stocks, which measure whether supply outweighs demand; and volatility of prices and the probability of high prices, which test the stability of global food security.

Adoption model

Diffusion is generally defined as the process by which a successful innovation gradually becomes broadly used through adoption by firms or individuals (Jaffe *et al.*, 2000). Most diffusion models of GM crops ignore the power of consumers because the direct agronomic benefits to farmers are assumed to be the main forces driving the diffusion. Most of the studies modelling diffusion of agricultural innovations claim that the variation of diffusion in different regions is explained by differences in 'farm factors' such as profitability or expected return from the adoption of the new technology (Griliches, 1957; Mansfield, 1961; Schultz, 1964; Dixon, 1980). However, these farm acceptance factors are related to how the final users of the farm product perceive the new technology. If consumers perceive the new technology as a risk for them or for the environment, they may not accept the farm product, thereby reducing the incentive for farmers to adopt the technology.

Recently, concerns of farmers in North America about possible introduction of GM wheat show that the strong consumer opposition in Europe and Japan to GM food cannot be ignored. In the literature, opposition to GM food, as reflected in consumer resistance, has not been recognized as an influencing factor in the adoption process. Rather, it is usually considered as a factor influencing demand for GM food, locally or internationally. Barkley (2002) estimated the opposition to GM food by a numerical factor $(-1, 0)$, reflecting demand reductions by foreign buyers. Anderson *et al.* (2001) also introduced opposition to GM food as a factor leading to a consumption ban on imports from GM-producing countries.

Our approach formally incorporates consumer response into the diffusion decision. Given that GM wheat is not yet available, we use a variable-slope dynamic diffusion model for GM maize to investigate the effect of consumer opposition on diffusion and adoption of GM technology. Using the available data about GM maize adoption,[1] we model the diffusion paths in 39 countries to identify the effect of local and foreign tolerance to GM food on the path of propagation of crop biotechnology.[2]

The variable-slope dynamic diffusion model for GM maize from 1996 to 2001 is presented by Equation 1:

$$\ln\left[\theta_{it}/(K_i - \theta_{it})\right] = a_i + \Phi_{it}t \tag{1}$$

The a_i parameter, a constant, measures the adoption at time $t = 0$, the year a new agricultural technology is first introduced (considered the base time period). Adoption behaviour is depicted by the percentage of total acreage planted with the GM crop. The adoption ceiling (K_i) is fixed for all countries at the 100% level. We take the coefficient of diffusion Φ_{it} as a function, $f(R_{it})$, where R_{it} is a vector of exogenous vectors affecting the diffusion coefficient. By defining $f(R_{it})$ as a function of foreign and national acceptance of GM food, $\Phi_1 R_{1it} + \Phi_2 R_{2it}$, this approach allows us to measure the diffusion of GM crops for different countries.

In order to capture some of the effects of consumer perceptions on rates of adoption, information on labelling requirements in each country was used to construct a tolerance index to GM food. The national tolerance index for country i (R_{1it}) is intended to be a proxy for the local acceptability of GM food:

$$R_{1it} = T_i \qquad (2)$$

The local tolerance level is taken to be inversely related to the stringency of the labelling requirements, i.e. countries with voluntary labelling laws or with no labelling regulation at all have a more tolerant attitude (higher T_i) toward GM food, whereas countries with mandatory regulation of 0% threshold on products containing GMOs are the least tolerant to GM food. The tolerance indexes are assigned to each country based on the threshold level for the percentage of GM contamination of non-GM products. Therefore, a value of 0 is assigned to countries with a 0% threshold, 1 for a 1% threshold, 2 for a 2% threshold, 3 for a 5% threshold, 4 for 10% and above, and 5 where either voluntary or no labelling laws exist. Although this method seems to be *ad hoc*, it is reasonable under the limitation of information on consumer preferences.

The foreign tolerance index facing country i (R_{2it}) is estimated as:

$$R_{2it} = \sum_j \frac{\text{Export}(i \text{ to } j)}{\text{Production}(i)} T_j \qquad (3)$$

where T_j is the tolerance level in importer country j approximated as mentioned above. The formula for R_{2it} suggests that country i will be affected the most by the labelling regulation (regarding a given product) of its most important trade partner. In other words, countries will take more seriously any labelling regulations or import restriction imposed on an export product coming from a large importer of that product. The ratio of export to total production is also taken into account in the analysis of the global diffusion of GM varieties. Crop export data and production data are obtained from FAOSTAT online statistics (Food and Agriculture Organization, 2000). Based on this information the following relationship is developed:

$$\Phi_{it} = \Phi_1 R_{1it} + \Phi_2 R_{2it} \qquad (4)$$

Substituting Φ_{it} in Equation 4 into the diffusion model 1 we get:

$$\ln\left[\theta_{it}/(K_i - \theta_{it})\right] = a_i + \left(\Phi_1 R_{1it} + \Phi_2 R_{2it}\right)t \qquad (5)$$

Country differences in biotechnology are accounted for by the term a_i which represents the fixed effect, as we are dealing with panel data. If at the initial year of introduction of the technology in the world ($t = 0$), a country does not adopt this technology, the term a_i will be negative. However,

if the country adopts the new technology at the initial year, the term a_i will be positive. Note that if the adoption rate (θ_{it}) is equal to zero the $\ln[\theta_{it}/(K_i - \theta_{it})]$ is invalid. To solve this problem, we substitute $\ln[\theta_{it}/(K_i - \theta_{it})]$ by $\ln[1 + \theta_{it}/(K_i - \theta_{it})]$. Equation 5 now becomes:

$$-\ln(1 - \theta_{it}) = \beta_i + (\alpha_1 R_{1it} + \alpha_2 R_{2it})t \qquad (6)$$

Adding the error term ε, we can estimate the diffusion equation as:

$$\Delta_t = -\ln(1 - \theta_{it}) = \beta_i + (\alpha_1 R_{1it} + \alpha_2 R_{2it})t + \varepsilon \qquad (7)$$

A negative intercept $\beta_i < 0$ means that there is no adoption at time zero in country i, while a positive value means the opposite. If diffusion is positive then the adoption rate in country i is equal to:

$$\theta_{it} = 1 - e^{-(\Delta t)} \quad \text{if } \Delta_t > 0 \quad \text{and} \qquad (8)$$

$$\theta_{it} = 0 \qquad\qquad \text{if } \Delta_t < 0 \qquad (9)$$

As mentioned for the term a_i, the term β_i is country specific, and is related to the initial availability or development of biotechnology in country i. We set this term as a function of field trials in previous years, in that field trials are one proxy of biotechnology research capacity at the national level.

It was expected that an increase in the magnitude of foreign and local tolerances would foretell an increase in the demand for genetically engineered crops, inducing wider adoption of the technology. In other words, it is expected that if GM technology is more accepted locally and internationally, adoption will increase globally. Consequently, both tolerance terms are expected to have positive coefficients.

The result of regression of the diffusion measure on the two acceptability variables is reported in Appendix 15.1. Both coefficients of acceptability variables have the right sign and are significant. Both foreign tolerance and local tolerance seem to perform well (significant at 1%). The ability of the two tolerance indices to account for differences in the diffusion parameters between countries appears to be significant, with 74% of the variability of the diffusion of GM crops worldwide explained by only these two variables.

The empirical results suggest that both tolerance measures affect the diffusion rate, but foreign acceptance of GM food is economically more significant than local tolerance. This is expected, since GM maize is adopted mainly by maize exporters. A developer would care about foreign opposition since the product is destined mainly for export markets. The significance of foreign tolerance suggests that foreign opposition to GM food could significantly limit the adoption of GM crops. This result helps to explain the hesitation of farmers to adopt GM wheat under the risk that they may lose export markets.

The coefficients β_i, α_1 and α_2 estimated in this section and presented in Appendix 15.1 (for GM maize) are used to forecast the diffusion of GM wheat for specific countries (Argentina, Australia, Canada, China and the USA). Foreign tolerance indices are re-calculated to reflect wheat trade shares rather than maize markets. Export over production shares used to calculate the foreign tolerances are 11 year averages (1990–2000) calculated from Canada Grains Council (2002).

The global adoption rate of GM wheat at time t is θb_t calculated as the acreage harvested with GM wheat over total wheat acreage. The global adoption of GM wheat is:

$$\theta b_t = \sum_1^k \frac{A_{it}\theta_{it}}{A_t} = \sum_1^k SA_{it}\theta_{it} \qquad (10)$$

where A_{it} is the acreage of wheat in country i (i.e. a potential adopter for GM wheat) at time t. The term θ_{it} is the adoption of GM wheat in country i at time t. A_t is the global acreage of wheat. SA_{it} is the acreage share of country i from the global acreage. Acreage shares used to calculate the global adoption are 10 year averages (1991–2000) calculated from Canada Grains Council (2002).

The market and food security elements of the model

The rest of the model is outlined here for completeness. The maximum obtainable yield can be estimated at the global level using conventional varieties (in 1999):

$$Max\,Yc_{1999} = (1 + \text{yield gap}_{1999})$$
$$\times Yc_{1999} = 3.30\,t/ha \qquad (11)$$

Assuming conventional technology can keep the growth rate (gc) between 2002 and 2030 at the level of annual yield growth in the period 1989–2002 (gc = 1.01%), the yield for conventional production at time t is:

$$Yc_t = Yc_{t-1}(1 + gc) \qquad (12)$$

The yield of conventional production in 2030 will be:

$$Yc_{2030} = Y_{2002}(1 + 0.0101)^{t-2002} = 3.57\,t/ha \qquad (13)$$

If conventional technology is able to keep the growth rate at 1.01% despite resource degradation, the conventional yield would be unable to reach 4 t/ha by 2042. The value of 4 t/ha is thus used as the maximum yield for the conventional technology in the future decades.[3] GM wheat may contribute to closing the yield gap from 2005 to T:

$$Yb_t = Yc_t + 0.1\,\delta 1\,(t - 2005)\text{Yield gap}_{2005} \qquad (14)$$

The constant 0.1 is an assumed annual decrease in the yield gap due to biotechnology. Here it is assumed that this decrease is a linear function of time. The term $\delta 1 = 0$ if the first generation of GM wheat is not developed or commercialized, and $\delta 1 = 1$ if it is commercialized. Before 2005, the term $\delta 1$ is equal to zero. Substituting Yc_t by its expression, we get:

$$Yb_t = Yc_{2005}(1 + 0.01)^{t-2005}$$
$$+ 0.1\,\delta 1\,(t - 2005)\text{Yield gap}_{2005} \qquad (15)$$

Solving this equation for the time (T) at which biotechnology will close the yield gap, we find that biotechnology closes the yield gap when

$$Yb_T = Max\,Yc_t \qquad (16)$$

Substituting Yb_t into Equation 15 yields:

$$Max\,Yc_t = Yc_{2005}(1 + 0.01)^{T-2005}$$
$$+ 0.1\,(T - 2005)\text{Yield gap}_{2005} \qquad (17)$$

Solving this equation for time (T), we find that the yield gap is closed in 2013. From 2013 to 2030, the effect of biotechnology on yield depends on whether any subsequent generations are developed and commercialized or not. Gray *et al.* (2001) suggest that the introduction of GM crops will attract private investment to the industry and accelerate the yield growth rate by 50%, such that the biotechnology yield beyond 2013 could be estimated by:

$$Yb_t = Yb_{2013}(1 + gc + \delta_2 xg)^{(t-2013)} \qquad (18)$$

where xg is the extra growth that biotechnology could offer. The term xg is assumed to be equal to (0.5 gc). The term δ_2 equals 0 if the second generation of GM wheat is not developed or commercialized; it equals 1 if GM wheat is developed and commercialized. Here it is assumed that once the second generation is commercialized, it will totally substitute for the first generation since it offers superior performances. Global yield is determined by:

$$Y_t = \theta b_t Y b_t + \theta c_t Y c_t = \theta b_t Y b_t + (1 - \theta b_t) Y c_t \tag{19}$$

where θb_t is the global adoption rate of GM wheat at time t, and θc_t is the global adoption rate of conventional wheat varieties at time t.

Yield data from the Food and Agriculture Organization (2004) database (FAOSTAT) are used to calculate yield variabilities. Under the assumption of no climate change, the random variables (v) are selected each year to create a pattern resembling that of the period 1989–2002. The expected variance is 1.6%, the variance that actually occurred over the period 1989–2002. To simulate greater variability under the assumption of climate change, the random variables are selected at each year to create a pattern resembling that of the period 1961–1974. Therefore, the variance under climate change assumption is equal to 3.85%, which is the yield variance in 1961–1974. Thus, the realized yield for the given year for each point j is:

$$Y_{rt} = Y_t(1 + v) \tag{20}$$

where v is a random variable generated subject to the expected variance.

The key parameters in the market model are the price elasticity of demand, price elasticity of area (or acreage response) and price elasticity of stock, which are based on other estimates. For the present study, the price elasticity of global consumption (or global demand) for wheat is chosen to be equal to -0.15.[4] In this study, the global acreage elasticity is assumed equal to 0.5.[5] The stock elasticity is calculated by dividing the demand elasticity by the average stocks to use ratio in the 1961–2000 period using United States Department of Agriculture data (in order to take into account the variability in stock to use ratio over the years), which yields a stock price elasticity of -0.55, indicating that storage is more responsive to price changes than consumption.

Global consumption at time t is determined by Equation 21. It is assumed that future global income growth will reduce the global consumption of wheat per capita (wheat is considered an inferior food) just enough to offset population increase. Therefore, the global consumption equation is:

$$GC_t = a + b \times P_t \tag{21}$$

The realized global production (PR_t) is the most important variable for supply. Global production is the product of wheat acreage (A_t) and realized yield (Y_{tr}).

$$A_t = c + d \times P_t \tag{22}$$

$$PR_t = A_t Y_{tr} \tag{23}$$

The total global supply (GS_t) of wheat is the sum of initial stock (S_{t-1}) and production.

$$GS_t = PR_t + S_{t-1} \tag{24}$$

where the term S_{t-1} is the ending global stock from the previous year. The global stock equation is calculated as:

$$S_t = e + f \times P_t \tag{25}$$

The equilibrium condition is satisfied when total demand is equal to total supply. Solving for the price to ensure the equilibrium will give us the real price at time t:

$$GC_t + S_t = GS_t \Rightarrow P_t = P_t^* \tag{26}$$

To check the effect of opposition to GM wheat on volatility of future prices, the expected prices – converted to inflation-adjusted terms using the US gross domestic product (GDP) price deflator index of 2000 – are arranged in two intervals. These intervals consist of low to medium expected prices (expected prices < 200 (i.e. US$2000/t)), and high expected prices (expected prices > 200 (i.e. US$2000/t)).

The base price for wheat (for the year 2000) is estimated to be around US$145/t.[6] Consumption, acreage and stock parameters are calibrated using this price. These specific calibrated parameters for the world are presented in Table 15.5.

This model is used below to simulate some scenarios of how food security is affected by various consumer responses to GM wheat. In each scenario, first local tolerance is assumed depending on each scenario (Equation 2). Secondly, the foreign tolerance index facing potential GM wheat adopters is calculated (Equation 3). Thirdly, the adoption rate in each of those countries is estimated (Equations 4–9). Fourthly, the global adoption rates of GM wheat are determined (Equation 10). The global adoption rate is then combined with the market model to compute the

Table 15.5. Calibrated consumption, acreage and stock parameters.

	Consumption parameters		Acreage parameters		Stock parameters	
	a	b	c	d	e	f
World	682.423	−0.614	107.135	0.739	176.4317	−0.433

Source: authors' calculations.

GFS element. It is important to emphasize that the goal from these scenarios is not the prediction of prices in the next decades. The main goal is the estimation of the impact that opposition to GM wheat will have on the expected price levels and variability compared with what would otherwise prevail.

Strategic Implications

This section presents a quantitative analysis using the GFS model developed above. The adoption model is shocked with different levels of EU tolerance to GM wheat while the supply model is shocked by different assumptions about how future GM wheat might affect yield growth along the GM wheat development path until the year 2030. In all scenarios, it is assumed that agricultural biotechnology will target genetic modification of yield potential in wheat. The scenarios are summarized in Table 15.6.

Scenario 1a involves the EU remaining non-GM even in the face of climate change. In this scenario, EU opposition induces widespread opposition to GM wheat in major importing countries. This widespread rejection of GM wheat would provide no incentive for GM food-producing countries to introduce GM wheat. Scenario 2a involves the EU accepting GM crops in response to climate change. In this scenario, EU consumers have maximum tolerance to GM wheat. If the EU has maximum tolerance to GM wheat, it is most likely that EU wheat farmers will adopt GM wheat in order to avoid lagging behind other exporting countries in wheat productivity. The EU stance (acceptance) vis-à-vis GM wheat will induce high consumer tolerance in major importing countries, since this European acceptance will be taken as endorsement of GM wheat and as an ending of the debate over the acceptability of GM food.

The other scenarios are identified as scenario 1c, 1d and 2d. The purpose of these additional scenarios is to create data points for comparison to allow the estimation of the impact of EU policy on food security. In scenario 1c, China adopts GM wheat despite the widespread rejection. China is not an exporter of wheat, therefore it may not be affected by this rejection. In scenario 1d, EU opposition induces only limited opposition to GM wheat in Magreb countries (Tunisia, Algeria and Morocco), Japan and South Korea. In scenario 2d, EU acceptance will induce widespread acceptance. However, in this scenario, Japan and South Korea remain opposed to GM wheat.

In all previous scenarios, it is assumed that climate change occurs. In order to account for the possibility of no climate change, the EU non-GM/GM scenarios 1a and 2a were rerun under an assumption of no climate change. Scenario 1b examines widespread opposition if climate change does not happen, while scenario 2b involves widespread acceptance under similar climatic conditions (Table 15.6).

It is important to note that 'Yes' for a major adopter or a major importer in Table 15.6 means total acceptance by that country, and the country is given a maximum tolerance (value of 5). Saying 'No' by an importer means that GM wheat is banned by that country and it is given a zero tolerance index. Saying 'No' by an adopter means that the GM wheat is not adopted, and that the tolerance to GM wheat is equal to zero. This tolerance scale (0–5) was introduced in the diffusion model previously discussed.

The scenarios are compared to estimate the impacts of the EU policy on global adoption of GM wheat and on global food security (Table 15.7).

Widespread induced opposition would cause the global adoption rate to decline over time (comparing scenarios 1a and 2a). Our results (Table 15.7, column 1) suggest that if the EU induces widespread opposition (whether climate change occurs or not), this will cause a decline in the GM adoption rate by as much as 25% by 2030. It is important to mention that the underlying

Table 15.6.   Opposition scenarios.

	EU Non-GM scenarios				EU GM scenarios		
	Scenario 1a	Scenario 1b	Scenario 1c	Scenario 1d	Scenario 2a	Scenario 2b	Scenario 2d
Climate change	Yes	No	Yes	Yes	Yes	No	Yes
Potential adopters: adopt or not							
Australia	No	No	No	Yes	Yes	Yes	Yes
Argentina	No	No	No	Yes	Yes	Yes	Yes
Canada	No	No	No	Yes	Yes	Yes	Yes
China	No	No	Yes	Yes	Yes	Yes	Yes
USA	No	No	No	Yes	Yes	Yes	Yes
EU	No	No	No	No	Yes	Yes	Yes
Major importers: import or not							
Magreb	No	No	No	No	Yes	Yes	Yes
Japan	No	No	No	No	Yes	Yes	No
South Korea	No	No	No	No	Yes	Yes	No
Others	No	No	No	Yes	Yes	Yes	Yes

Table 15.7.   Effect of opposition to GM wheat on adoption and food security elements (%).

	Widespread opposition with climate change	Widespread opposition with no climate change	Limited opposition with climate change	Japan and South Korea opposition	China adopts	Effect of climate change
Adoption rate	−25	−25	−5	−1	4	0
Expected price	+16.33	+17.01	+2.94	+0.46	−2.30	+4.96
Availability	−1.81	−2.02	−0.32	−0.05	0.30	−0.52
Consumption	−1.29	−1.28	−0.23	−0.04	0.21	−0.37
Ending stocks	−4.10	−5.32	−0.74	−0.12	0.70	−1.17
Probability of high prices	0.00	0.00	0.00	0.00	0.00	1.00

assumption in the adoption model – that only the major wheat exporters and China are likely to adopt GM wheat in any circumstances – leads to an underestimate of the effect of widespread opposition to global adoption. The effect of widespread opposition varies depending on whether climate change occurs. If climate change occurs, widespread opposition would deprive the world of a range of potentially more productive varieties (GM wheat). Therefore, overall potential availability declines by as much as 1.8%, causing the

world price for wheat to be more than 16% higher in 2030 than it otherwise might be. The increased world expected prices would cause consumption to decline by about 1.3% in 2030. Widespread opposition would negatively affect the ending stocks, by up to 4% in 2030. It is important to mention that consumption of wheat in poor countries is more sensitive to expected price changes, and therefore it is expected that most of the consumption decline due to expected price increase would be in poor countries. If climate change does not occur and there is widespread opposition (comparing scenarios 1b and 2b, column 2, Table 15.7), the results are slightly different. Widespread opposition would lower availability by 2%, leading to an expected price rise of 17% in 2030. This increase in expected price would reduce consumption by 1.3% in the year 2030. As a result of this expected price increase, the ending stocks would shrink by around 5% in 2030. The probability of high expected prices is not predicted to increase in 2025–2030 because of widespread opposition whether climate change occurs or not.

If the EU induces limited opposition to GM wheat (comparing scenarios 1d and 2a, column 3, Table 15.7), there would be only a small global adoption reaction (5% in 2030). The decline in global adoption due to limited rejection would decrease wheat availability by not more than 0.3%. This decline in availability, although small, would lead expected prices to increase by around 3% in 2030. The expected price increase would depress consumption by only 0.2%. Limited opposition is not expected to change the probability of high expected prices.

If Japan and South Korea, in contrast, oppose the technology (comparing scenarios 2a and 2d, column 4, Table 15.7), there would only be a very small influence on the availability of wheat. The greatest impact this opposition might have would be to decrease the availability by 0.1% in 2030. As a result, the effect of the opposition of Japan and South Korea on expected prices does not exceed 0.5%. This is expected to have a very small effect on the future consumption. This opposition has no effect on the probability of high expected prices.

There is a real possibility that China might adopt GM wheat, regardless of what others do. If China rapidly adopts GM wheat (comparing scenario 1a and 1c, column 5, Table 15.7), global adoption would rise 4% by 2030. This would lead to an increase in global wheat availability, thereby lowering expected prices by 2.3% in 2030. The expected price decline would lead to consumption increasing by 0.2% in 2030. The ending stocks would increase by 0.7% in 2030 (Table 15.7). China's adoption would have no effect on the volatility of expected prices, since it has no effect on probability of high expected prices. Relatively speaking, China's decision about GM wheat will have a more significant impact on global food security than the impact of either the Magreb countries or Japan and South Korea.

In the diffusion model, climate change was not included as a driving factor – climate change is assumed to neither motivate nor discourage countries to adopt GM wheat. However, climate change is expected to affect availability, expected price, consumption and ending stock by affecting the variation of yields (comparing scenarios 2a and 2b, column 6, Table 15.7). Climate change by itself will decrease the availability of wheat by as much as 0.5% in 2030, which would lead to around a 5% increase in expected prices. The consumption reaction to this expected price increase would be −0.37%, and the ending stocks would decline by more than 1%. Perhaps most importantly, climate change is expected to significantly increase the probability of high expected prices in the coming decades, which would particularly imperil poor consumers in developing countries.

Conclusions

The most important finding of this study is that opposition to GM wheat could exacerbate GFS (measured by price, availability, consumption and ending stocks). The magnitudes of these losses are directly related to the degree of opposition to GM wheat. If the EU induces widespread opposition (whether climate change occurs or not), this will greatly diminish GFS since it deprives the world of the benefits that GM wheat may offer. Widespread opposition to GM wheat would lead to a 25% decline in the global adoption of this technology in 2030, which would lead to a 16% increase in expected prices by 2030. The price increase leads to a decline in consumption and ending stocks, and ultimately more hungry people.

However, if the EU induces only limited opposition by affecting the position of only a few countries, such as the Magreb countries (Tunisia, Algeria and Morocco) or Japan and South Korea, this will not have a significant effect on global food security. The limited opposition to GM wheat would lead only to a 5% reduction in global adoption reaction in 2030, which would have a very small effect on availability, price, consumption and ending stocks.

It is interesting to note that the boost to global food security if China adopted GM wheat could almost totally offset the effect of limited opposition elsewhere. Driven by its local policy only, China's adoption could rise to 30% by 2030, which would depress wheat prices by around 3% in 2030.

The results of this research are important in terms of future policy and regulations regarding release as well as trade of GM products. Many countries still oppose GM crops. GM-exporting countries should not ignore those concerns. We have learned from recent bovine spongiform encephalitis (BSE) and bird influenza outbreaks that consumer perception of safety of a given product can fundamentally alter the trade of those products. It is obvious that consumer concern can be the main driver as well as the main obstacle to adoption. Regardless of the reasons or concerns behind this resistance, those concerns should be taken into account in the development and evaluation of new products before their release and commercial development. Opposing (consumers) countries should similarly recognize that their opposition has the potential to deprive the world of more food.

Consumer concern over GM food is an issue that needs attention, whether this opposition is based on science or not. It is urgent that opponents and proponents of GM food arrive at a midpoint that satisfies both parties. The failure to do so may deprive the world of a technology that could contribute to fewer hungry people. It is understandable that interests of different parties over the issue of GM food may conflict. However, all parties should consider food security as an important component in the international trade negotiations over GM food.

Notes

[1] The analysis is conducted using a sample of 39 countries for which biotechnology research statistics are available, and covers 5 years of adoption of GM maize (1996–2001).

[2] Adoption levels for GM maize are taken from various issues of Clive James, *Global Review of Commercialized Transgenic Crops: ISAAA Briefs* (International Service for the Acquisition of Agri-biotech Applications: Ithaca, New York, 1995–2000).

[3] This value is a long run upper limit and not technical maximum yield.

[4] The overall demand for wheat is regarded as highly inelastic. According to Alston *et al.* (1997), the elasticity of demand for food wheat is equal to –0.2. In this study, the results did not change for the elasticity of demand in the range of –0.1 and –0.2. This range is similar to the one that Alston

et al. (1994) assumed for the overall elasticity of demand for milling wheat and durum wheat. Sullivan *et al.* (1989) reported price elasticity of demand for major importers ranging from –0.1 to –0.4, with the rest of the world having a demand elasticity equal to –0.25. Benirshka and Koo (1995) determine elasticities of demand for different countries and regions. The elasticity values vary from –0.005 for Japan to –0.3 for Australia. DeVuyst *et al.* (2001) calculates the demand elasticity for the rest of the world from this data and found it equal to –0.1.

[5] The intermediate run value of acreage response was estimated by Alston *et al.* (1994) to be equal to 0.5 (or less) for Canada and the rest of the world. The Organization for Economic Cooperation and Development (1986) estimates the supply elasticity to be 0.5 for the USA and 0.46 for the EU.

[6] Estimated from average US$-denominated free on board Gulf of Mexico export price for US HRS No. 2.

References

Alston, J.M., Gray, R. and Sumner, D.A. (1994) The wheat war of 1994. *Canadian Journal of Agricultural Economics* 42, 231–251.

Alston, J.M., Carter, C.A. Gray, R. and Sumner, D.A. (1997) Third-country effects and second-best grain trade policies: export subsidies and bilateral liberalization. *American Journal of Agricultural Economics* 79, 1300–1310.

Anderson, K., Nielsen, C.P., Robinson, S. and Thierfelder, K. (2001) Estimating the global economic effects of GMOs. In: Pardey, P.G. (eds) *The Future of Food: Biotechnology Markets and Policies in an International Setting.* International Food Policy Research Institute, Washington, DC, pp. 49–74.

Angelo, I., Masiga, F. and Musiita, L. (2003) Africa's dilemma in genetically modified food. war.allAfrica.com. Kampala.

Barkley, A.P. (2002) *The Economic Impacts of Agricultural Biotechnology on International Trade, Consumers, and Producers: the Case of Corn and Soybeans in the USA.* In: 6th International ICABR Conference. Ravello, Italy.

BBC (2002) Zambia turns down GM aid. *BBC News*, 17 August 2002.

Benirshka, M. and Koo, W.W. (1995) *World Wheat Policy Simulation Model: Descriptive and Computer Program Documentation.* Agricultural Economics Report No. 340, Department of Agricultural Economics, North Dakota State University, Fargo, North Dakota.

Byerlee, D. and Moya, P. (1993) *Impacts of International Wheat Breeding Research in the Developing World, 1969–1990.* CIMMYT, Mexico, DF.

Canada Grains Council (CGC) (2002) *Canadian Grains Industry Statistical Handbook.* Winnipeg, Manitoba.

Canadian Wheat Board (2003) *Current State of Market Acceptance and Non-acceptance of GM Wheat.* Winnipeg, Manitoba.

Cassman, K. and Pingali, P. (1995) Extrapolating trends from long-term experiments to farmers' fields: the case of irrigated rice systems in Asia. In: Barnett, V. Payne, R. and Steiner, R. (eds) *Agricultural Sustainability: Economic, Environmental and Statistical Considerations.* Wiley, New York, pp. 63–84.

CIMMYT (1996) *World Wheat Facts and Trends 1995/96: Understanding Global Trends in the Use of Wheat Diversity and International Flows of Wheat Genetic Resources.* International Maize and Wheat Improvement Center (CIMMYT), Mexico, DF.

Cropchoice (2003) GM wheat seen getting mixed reception in Asia. *Cropchoice News*, 17 June, 2003.

Daily Times (2003) Pakistan Feels EU Labeling Pressure. *Daily Times*, 25 August, 2003.

DeVuyst, E.A., Koo, W.W., DeVuyst, C.S. and Taylor, R.D. (2001) *Modeling International Trade Impacts of Genetically Modified Wheat Introductions.* Agribusiness and Applied Economics Report No. 463, Department of Agribusiness and Applied Economics, North Dakota State University, Fargo, North Dakota.

Dixon, R. (1980) Hybrid corn revisited. *Econometrica* 48, 1451–1461.

Food and Agriculture Organization (1985) *Report of the Tenth Session of the Committee on World Food Security.* FAO, Rome.

Food and Agriculture Organization (1996) *World Food Summit. Rome Declaration on World Food Security and World Food Summit Plan of Action.* FAO, Rome.

Food and Agriculture Organization (1998) *Regional Development Partnerships Programme (RDPP).* FAO, Rome.

Food and Agriculture Organization (2002) *The State of Food Insecurity in the World 2002: Agriculture and Global Public Goods Ten Years After the Earth Summit.* FAO, Rome.

Food and Agriculture Organization (2004) *FAOSTAT Database.* FAO, Rome.

Gray, R., Malla, S. and Ferguson, S. (2001) Agriculture Research Policy for Crop Improvement in Western Canada: Past Experience and Future Directions. Unpublished manuscript, Centre for Studies in Agriculture, Law and the Environment, University of Saskatchewan, Saskatchewan.

Griliches, Z. (1957) Hybrid corn: an exploration in the economics of technological change. *Econometrica* 25, 501–522.

Huang, J., and Rozelle, S.D. (1995) Environmental stress and grain yields in China. *American Journal of Agricultural Economics* 77, 246–256.

Jaffe, A.B., Newell, R.G. and Stavins, R.G. (2000) *Technological Change and the Environment.* Resources for the Future, Discussion Paper 00-47, Washington, DC.

Lacy, W. and Busch, L. (1986) Food security in the United States: myth or reality. In: Subcommittee on Agriculture and Transportation, Joint Economic Committee, Congress of the United States. *New Dimensions in Rural Policy: Building Upon Our Heritage.* US Government Printing Office. Washington, DC, pp. 222–223.

Luce, E. (2003) India rejects US food shipment over GM content. *The Financial Times.* New Delhi, India.

Mansfield, E. (1961) Technical change and the rate of imitation. *Econometrica* 29, 741–766.

Organization for Economic Cooperation and Development (1986) *Update of Elasticities Tables for the MTM Model.* Paris.

Pingali, P.L. and Heisey, P.W. (2001) Cereal-crop productivity in developing countries: past trends and future prospects. In: Alston, J.M., Pardey, P.G. and Taylor, M.J. (eds) *Agricultural Science Policy: Changing Global Agendas.* The Johns Hopkins University Press, Baltimore, Maryland.

Pinstrup-Andersen, P. and Schioler, E. (2001) *Seeds of Contention: World Hunger and the Global Controversy over GM (Genetically Modified) Crops.* The Johns Hopkins University Press, Baltimore, Maryland.

Pinstrup-Anderson, P., Pandya-Lorch, R. and Rosegrant, M.W. (1999) *World Food Prospects Critical Issues for the Early Twenty-First Century.* Food Policy Statement No. 29. International Food Policy Research Institute, Washington, DC.

Rosegrant, M.W., Sombilla, M.A., Gerpacio, R.V. and Ringler, C. (1997) Global food markets and US exports in the twenty-first century. In: *World Food and Sustainable Agriculture Program Conference, Meeting the Demand for Food in the 21st Century: Challenges and Opportunities.* University of Illinois, Urbana-Champaign, Illinois.

Sayre, K.D. (1996) The role of crop management research at CIMMYT in addressing bread wheat yield potential issues. In: Reynolds, M.P., Rajaram, S. and McNab, A. (eds) *Increasing*

Yield Potential in Wheat: Breaking the Barriers, CIMMYT Mexico, DF, pp. 203–207.
Schultz, T.W. (1964) *Transforming Traditional Agriculture*. Yale University Press, New Haven, Connecticut.
Sullivan, J., Wainio, J. and Roningen, V.O. (1989) *A Data Base for Trade Liberalization Studies*. US Department of Agriculture, ERS Staff Report No. 89. Washington, DC.
The Economist (2003) Genetically Modified Food. April 3, 2003.

US Wheat Associates (2002) *GM Wheat Customer Acceptability Survey – Results from Asia*.
Weiss, R. (2002) Zimbabwe continues to block gene-altered corn. *Washington Post,* p. A14.
World Bank (1986) *Poverty and Hunger: Issues and Options for Food Security in Developing Countries*. World Bank, Washington, DC.
World Bank (1992) *World Development Report 1992. Development and the Environment*. World Bank, Washington, DC.

Appendix 15.1. Estimation of the Diffusion Parameters (Case of Maize)

Dependent Variable: Diffusion of GM Crop
Method: Pooled Least Squares
Sample: 1996 2000
Included observations: 5
Number of cross-sections used: 24
Total panel (balanced) observations: 120

Variable	Coefficient	Std. Error	t-Statistic	Prob.
Local Tolerance *t	0.003	0.0010	2.73	0.0076
Foreign Tolerance *t	0.013	0.0058	2.21	0.0296

Fixed Effects

EU (15)	−0.006
Argentina	−0.038
Australia	−0.036
Brazil	−0.017
Bulgaria	−0.019
Canada	**0.300**
China	−0.024
Czech Republic	−0.013
Hungary	−0.045
India	−0.026
Israel	−0.018
Japan	−0.022
Korea	−0.024
Liechtenstein	−0.026
Mexico	−0.021
New Zealand	−0.018
Norway	−0.015
Poland	−0.026
Romania	−0.018
Russian Federation	−0.021
Slovakia	−0.040
South Africa	−0.016
Turkey	−0.026
USA	**0.211**

R-squared	0.74	Mean dependent var	0.0281
Adjusted R-squared	0.68	SD dependent var	0.0989
SE of regression	0.06	Sum squared resid	0.2984
F-statistic	272.84	Durbin-Watson stat	1.0911
Prob (F-statistic)	0.00		

16 International Impacts of Bt Cotton Adoption

George B. Frisvold[1], Russell Tronstad[1] and Jeanne M. Reeves[2]

[1]Department of Agricultural and Resource Economics, University of Arizona, Tucson, Arizona; [2]Agricultural Research Department, Cotton Incorporated, Cary, North Carolina, USA

Introduction

In 2001, roughly 4 Mha of cotton were planted to Bt cotton. This includes Bt-only varieties and stacked Bt and herbicide-tolerant varieties. With nearly 2.4 Mha of Bt cotton planted in 2001, the USA accounted for 60% of world Bt cotton acreage (Williams). China planted nearly 1.5 Mha (James, 2001; Huang *et al.*, 2002b) and Australia roughly 0.1 Mha (Cotton Research and Development Council, 2002). Smaller areas of Bt cotton were also planted in Argentina, Indonesia, Mexico and South Africa (James, 2001; Ismael *et al.*, 2002; Qaim and de Janvry, 2002). A growing body of literature reports that Bt cotton has led to significant yield gains, reductions in conventional insecticide sprays or both throughout the world (Edge *et al.*, 2001; Pray *et al.*, 2001; Price *et al.*, 2001; Cotton Research and Development Council, 2002; Doyle *et al.*, 2002; Frisvold and Tronstad, 2002; Huang *et al.*, 2002a, b, c; Ismael *et al.*, 2002; Qaim and de Janvry, 2002; Traxler *et al.*, 2002; Qaim and Zilberman, 2003). These studies examine adoption impacts in one region, in isolation of adoption impacts in others. In general, they do not examine impacts of Bt cotton adoption on world or domestic cotton prices.

This study develops a three-region, output price endogenous model of the world cotton market to evaluate the global impacts of Bt cotton adoption in the USA and China in 2001. These two countries accounted for roughly 40% of world cotton production and over 95% of Bt cotton production in 2001. Although modest adoption occurred in the third, rest of world region, these impacts would be small on a world scale. In this study, the rest of the world is affected by Bt cotton adoption only via changes in the world price of cotton.

Model Structure

We assume a 'putty-clay' production technology. Acreage decisions and technology choice are flexible at the beginning of the crop year. Once growers make planting and initial technology decisions, however, production is characterized by a fixed proportion technology. At the beginning of the crop year, growers in each region i choose how much cotton to plant A_i, subject to a capacity constraint $\bar{A}_i$, which limits the total acreage where the crop may be grown profitably. Growers allocate cotton acreage between conventional cotton acres A_{i1} and Bt cotton acres A_{i2}, where $\bar{A}_i \geq A_{i1} + A_{i2}$. Growers allocate cotton acreage between Bt and conventional seed varieties to maximize expected profits.

Bt seed varieties command a price premium. This was originally embodied in a per acre technology fee and is now included in a per bag price premium. Bt varieties, however, reduce the need for conventional insecticide applications to control cotton bollworm, pink bollworm and tobacco budworm. To the extent that they control pest damage, Bt varieties exhibit higher yields.

Growers choose conventional and Bt cotton acreage (A_{i1} and A_{i2}) to maximize profits subject to constraints. The first is the overall capacity constraint, limiting total cotton acreage planted in a region, $\overline{A}_i$. Given the Leontief constant returns technology, if a producer can earn profits from producing using either technology, the overall capacity constraint will be binding.

In a given region and year, Bt cotton may be more profitable than conventional cotton on only a subset of potential cotton acreage. Within a region, there will be areas where bollworm/budworm pressure neither exceeds insecticide treatment thresholds nor causes appreciable yield losses. For example, in 1995, 15% of US cotton acreage was not infested by bollworm/budworms, while 37% of acreage did not receive insecticide treatments for these pests (Williams, 1996). On these acres, there is little scope for Bt cotton to reduce pest control costs or increase yields. To capture this relationship, we introduce a second land use constraint – an adoption ceiling, $\overline{A}_{i2}$. The ceiling reflects the fact that Bt cotton may have a profit advantage over conventional cotton on part, but not all, of a region's potential cotton acreage.

Another factor limiting adoption is regulation. To slow the development of pest resistance to Bt cotton, the Environmental Protection Agency (EPA) requires Bt cotton adopters to plant refuges of conventional cotton to allow susceptible and resistant pests to interbreed. Refuges must be at least 20% of cotton acreage if conventional sprays are used on the refuge to control target pests, and 5% of cotton acreage if the refuge is not sprayed for target pests.

The Lagrangian, L_i, for profit maximization in region i is:

$$L_i = (P_i + S_i)(A_{i1}Y_{i1} + A_{i2}Y_{i2})$$
$$- \{A_{i1}Y_{i1}[(C_{i1}/Y_{i1}) + G_i] + A_{i2}Y_{i2}$$
$$\times [(C_{i2}/Y_{i2}) + G_i]\} + \lambda_i(\overline{A}_i - A_{i1} - A_{i2})$$
$$+ \gamma_i(\overline{A}_{i2} - A_{i2}) \tag{1}$$

Yields and per acre costs for conventional cotton are Y_{i1} and C_{i1}, while for Bt cotton they are Y_{i2} and C_{i2}. For each region, yields and costs are constant for each technology within a given crop year. Per pound ginning costs (G_i) are the same for each technology. Producers select conventional cotton acreage (A_{i1}) and Bt cotton acreage (A_{i2}) to maximize profits, subject to the land capacity constraint

and the Bt cotton adoption ceiling. For every pound of lint produced, growers receive the market price P_i and a government support price payment, S_i. The terms λ_i and γ_i are the shadow costs of the land use constraints.

The market price (P_i) a grower receives for a pound of cotton lint is:

$$P_i = P_f + z_i \tag{2}$$

where P_f is the average US farm price and z_i is the regional price premium or discount that reflects difference in lint quality and transportation costs. The US farm price of cotton is a function of the endogenously determined world price, P_w

$$P_f = \theta P_w^\varepsilon \tag{3}$$

where θ and ε are parameters. Domestic and world prices can differ because of quality difference, transportation costs and government market interventions. The term ε is a price transmission elasticity that determines the percentage change in the US farm price in response to world price changes. The transmission elasticity captures the extent to which domestic price is sheltered from changes in world price. Following Sullivan *et al.* (1989), we assume $\varepsilon = 1$.

The government price support payment rate (S_i) is:

$$S_i = l_i P_w \tag{4}$$

where l_i is the weighted average of the loan deficiency payment rate and marketing gain payment rate per pound. The programme payment rate (l_i) is determined by the difference between the adjusted world price ($P_w - \omega$) and the loan rate, R

$$l_i = \max [0, R - (P_w - \omega)] \tag{5}$$

Producers receive payments if the adjusted world price falls below the loan rate, which is set at 51.92 cents per pound. The adjustment factor ω is set by the United States Department of Agriculture (USDA) and is based on transport cost and cotton grade considerations. It typically ranges from 12 to 14 cents. Payment rates vary by region, ranging between 25 and 28 cents per pound in 2001.

The solution to the Lagrangian yields the optimal allocation of acreage to conventional and Bt cotton as functions of the regional market price received and programme payment rates, $A_{i1}^*(P_i, S_i)$ and $A_{i2}^*(P_i, S_i)$. From Equations 2–5, optimal regional acreage allocations can be expressed as functions of the world price of cotton, $A_{i1}^*(P_w)$ and $A_{i2}^*(P_w)$. The regional supply of cotton Q_i^S is

$$Q_i^S = A_{i1}^*(P_w)Y_{i1} + A_{i2}^*(P_w)Y_{i2} \qquad (6)$$

There are 28 US regions in the model. These are Alabama-Central, Alabama-North, Alabama-South, Arizona, Arkansas-Northeast, Arkansas-Southeast, California-Imperial Valley, California-San Joaquin Valley, Florida, Georgia, Louisiana, Mississippi-Delta, Mississippi-Hills, Missouri, New Mexico, North Carolina, Oklahoma, South Carolina, Tennessee, Texas-Coastal Bend, Texas-Far West, Texas-High Plains, Texas-Lower Rio Grande Valley, Texas-North Central (N. Blacklands), Texas-North Rolling Plains, Texas- South Central (S. Blacklands), Texas-Southern Rolling Plains and Virginia.

The US supply of cotton lint is the sum of optimal production over all 28 production regions in the model

$$Q_u^S = \sum_{i=1}^{28} A_{i1}^*(P_w)Y_{i1} + A_{i2}^*(P_w)Y_{i2} \qquad (7)$$

Equation 7 generates a step–function supply curve, where each step represents marginal costs and production of cotton in each region for each technology (Fig. 16.1).

The supply functions for China (Q_c^S) and the rest of the world (Q_r^S) are linear

$$Q_c^S = \alpha_c(1+k) + \beta_c P_w \qquad (8)$$
$$Q_r^S = \alpha_r + \beta_r P_w \qquad (9)$$

where the α and β terms are scalar constants. To estimate the impact of China's adoption of Bt cotton, a supply shift parameter, k, is introduced into China's supply function. Yield increases and cost reductions from Bt cotton adoption are reflected in the size of k. A large value of k implies that more cotton will be supplied at any given market price.

Domestic US demand Q_u^D, Chinese demand Q_c^D and rest of the world demand Q_r^D are linear functions:

$$Q_u^D = a_u - b_u\theta P_w \qquad (10)$$
$$Q_c^D = a_c - b_c P_w \qquad (11)$$
$$Q_r^D = a_r - b_r P_w \qquad (12)$$

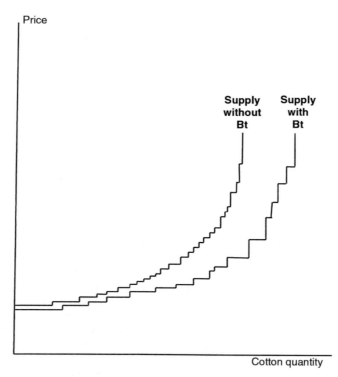

Fig. 16.1. Impact of Bt cotton on US cotton supply.

where the a and b terms are scalar constants. US supply equals domestic plus export demand:

$$Q_u^S = Q_u^D + Q_u^E \tag{13}$$

Export demand can be expressed as a function of the world price of cotton. The model is calibrated to 2001 data. In that year, the rest of the world was a large importer of cotton. China was also a net importer, but imports were small, only about 1% of consumption. The trade balance equation requires that net exports equal net imports:

$$Q_u^E = (Q_c^D - Q_c^S) + (Q_r^D - Q_r^S) \tag{14}$$

where Q_u^E is US exports. The US export demand equation can be expressed as a function of the world cotton price and exogenous variables by substituting Equations 8, 9, 11 and 12 into Equation 14:

$$Q_u^E = [(a_c + a_r) - (\alpha_c(1+k) + \alpha_r)] \\ -P_w(b_c + b_r + \beta_c + \beta_r) \tag{15}$$

The equilibrium world price can be determined by substituting Equations 7, 10 and 14 into Equation 13 and solving for P_w

$$\sum_{i=1}^{28} A_{i1}^*(P_w)Y_{i1} + A_{i2}^*(P_w)Y_{i2}$$

$$= [(a_u + a_c + a_r) - (\alpha_c(1+k) + \alpha_r)] \\ - [P_w(b_c + b_r + b_u\theta + \beta_c + \beta_r)] \tag{16}$$

The equilibrium world price is the P_w that solves Equation 16.

Model Calibration and Data

The model is calibrated so that the equilibrium solution replicates observed conventional and Bt cotton acreage, average yields, pest control costs, cotton prices, Bt adoption costs and cotton programme payments for the base year 2001 in each US production region. The baseline model also replicates the world cotton price as well as aggregate production, consumption, imports and exports for the USA, China and the rest of the world.

US regional and aggregate data sources used in the model are discussed in Frisvold *et al.* (2000) and Frisvold and Tronstad (2002). Estimates of

domestic and export demand elasticities were based on Isengildina *et al.* (2000), Meyer (1999), Price *et al.* (2001) and Sullivan *et al.* (1989). The consumption, production and demand for US exports by the rest of the world were derived from the Production Estimates and Crop Assessment Division of USDA's Foreign Agricultural Service, the International Cotton Advisory Council (ICAC) and from various issues of the USDA Economic Research Service *Cotton and Wool: Situation and Outlook Yearbook*. The 28 regions within the USA correspond to those reported in the *Cotton Crop Loss* database (Williams), with the addition of a Southern California region.

Modelling Supply Shocks from Bt Cotton Adoption

In the baseline model, US acreage, yields, prices, programme payment rates, exports and costs are calibrated to actual USDA data. Cotton production, consumption, demand for US cotton exports and the world price of cotton of China and the rest of the world are also set equal to USDA and cotton industry data. Implicitly, these data already account for the impacts of US and Chinese Bt cotton adoption.

To estimate the impact of Bt cotton adoption, we ask the counterfactual question, 'What would the US and Chinese cotton supply functions look like if Bt cotton had not been adopted?' For the USA, the programming model is constrained so producers can only grow conventional cotton. The impacts of US Bt cotton adoption are measured by the differences between the baseline and constrained models. This approach is similar to previous analyses of pesticide cancellations (Deepak *et al.*, 1996; Sunding, 1996). The effect of Bt cotton adoption on the US supply step function is shown in Fig. 16.1.

Estimates of changes in US insecticide application rates from Bt cotton adoption were derived from surveys of empirical studies of Bt cotton adoption impacts (Carpenter and Gianessi, 2001; Marra, 2001; Gianessi *et al.*, 2002). Data on costs per insecticide treatment for target pests and Bt cotton technology fees were obtained from the *Cotton Crop Loss* database. Individual regional impacts were aggregated to obtain a national supply shift. In the simulations, Bt cotton adoption reduced insecticide use by a weighted-average of 2.4 applications per acre in the USA. In a survey of empirical studies, Marra (2001) reported large variations in the impact of Bt cotton on insecticide use. In most major Bt cotton-adopting areas, however, the mean impact was a reduction of between two and three applications per acre. So, our simulation assumptions seem in line with this overall finding. While Bt cotton reduces costs for insecticide applications, adopters must pay higher prices for Bt seed. In our simulations, US insecticide cost savings exceeded higher seed costs by only US$1 million. This came out to a net pest control cost saving of only US$0.23 per acre. Gianessi *et al.* (2002) estimated that Bt cotton actually led to a US$2.00 per acre *increase* in net pest control costs, with technology fees exceeding insecticide cost savings.

In either case, it is not per acre cost savings that are the major economic incentive for adopting Bt cotton, but yield gains. To estimate the impacts of Bt cotton adoption on US producer yields, percentage yield increases were taken from the 'moderate impact scenario' developed in Frisvold *et al.* (2000) and Frisvold and Tronstad (2002). While, in the simulations, Bt cotton had little impact on costs per acre, it had more of an impact on cost per pound as yield increased. In the simulations, total US cotton production was 2.7% greater with Bt cotton adoption.

If Bt cotton were not adopted in China, the China supply function would shift upward in a parallel fashion. Recall the supply function (Equation 8) is $Q_c^S = \alpha_c(1+k) + \beta_c P_w$ with $k = 0$ in the baseline model. Without Bt cotton, k becomes negative. This is similar to the approach used by Lichtenberg *et al.* (1988) to estimate impacts of pesticide cancellations. Through the market equilibrium equation, these shifts induce a shift in the equilibrium world price of cotton. One can then simulate how much higher the world price would have been had there been no US or Chinese Bt cotton adoption.

To construct estimates of the k-shift parameter, we rely on information and data provided in Huang *et al.* (2002b) for China. Bt cotton accounted for 31% of total cotton acreage (Pray *et al.*, 2002). Several econometric studies have examined the farm-level impact of Bt cotton adoption on Chinese cotton production costs and yields (Pray *et al.*, 2001, 2002; Huang *et al.*, 2002a, c). Based on these studies, Huang *et al.* (2002b) reported a yield advantage of Bt cotton of 8.3% in

Hebei, Henan and Shandong provinces. These provinces accounted for 86% of Bt cotton acreage and 43% of all cotton acreage in China in 2001. The yield advantage in Anhui, Jiangsu and Hubei provinces was 5.8%. These provinces accounted for 12 and 24% of Bt cotton and total cotton acreage. The reported yield advantage in the remainder of China was 3%. This area accounted for a third of total cotton acres, but only 2.5% of Bt cotton acres. Based on these numbers, we assumed that Bt cotton adoption in China shifted supply in such a way to increase production 2% (at baseline price) and to reduce the marginal cost of production (at baseline quantity) by 24%. These assumptions appear in line with other studies (Pray *et al.*, 2001; Huang *et al.*, 2002b, c).

Impacts on Price, Production, Consumption and Trade

We consider three scenarios: (i) Bt cotton adoption in the USA only; (ii) adoption in China only; and (iii) adoption in both the USA and China. Bt cotton increases cotton production and consumption and reduces world and US prices. The effects are greatest with adoption in both areas followed by adoption in the USA alone, then China alone (Table 16.1). Under joint US–Chinese adoption, increased production contributed to a 1.4 cents per pound reduction in the world price of cotton.

In all scenarios, the rest of the world increases consumption, reduces production and increases its cotton imports, with the effects stronger with combined adoption. China's production increases (and imports decrease) the most if it is the sole adopter, but production declines (and imports increase) if the USA is the sole adopter. US exports rise 4.3% if it is the sole adopter. With adoption also occurring in China, US exports increase by only 3.6%. If adoption occurred only in China, US exports would fall by 0.7%.

Welfare Impacts

The changes in economic welfare in China and the rest of the world are measured as the sum of changes in producer and consumer surplus (Table 16.2). For the USA, the change in welfare

Table 16.1. Impact of Bt cotton adoption on cotton prices, production, consumption and trade, 2001.

	Bt cotton adoption in		
	USA only	China only	Both the USA and China
Change in US cents per pound			
World price	−0.7	−0.7	−1.4
US LDP[a] rate	0.7	0.7	1.4
US farm price received	−0.6	−0.5	−1.1
Percentage change			
US consumption	0.9	0.8	1.7
US production	2.7	0.0	2.7
US exports	4.3	−0.7	3.6
Chinese consumption	0.6	0.5	1.1
Chinese production	−0.3	1.8	1.5
Chinese imports[b]	60.2	−82.9	−22.7
ROW[c] consumption	0.2	0.2	0.4
ROW production	−0.2	−0.2	−0.3
ROW imports	2.3	2.2	4.5
World cotton production	0.4	0.4	0.7

[a]LDP = loan deficiency payment.
[b]In 2001, net imports of upland cotton in China were small, about 1% of consumption. Large percentage changes represent small changes in import levels.
[c]ROW = rest of the world.

is measured as the sum of the change in US producer surplus, consumer surplus and innovator-monopolist rents charged to US producers for seed, minus the change in US government programme payments to cotton producers. Including innovator-monopolist rents follows the approach introduced by Moschini and Lapan (1997). Ideally, one would also want to include measures of these rents captured in China. There, Bt seed varieties are supplied both by Monsanto/Delta and Pine Land and the Chinese Academy of Agricultural Science (CAAS). At the time of writing, we did not have access to information about any monopoly rents captured in China in 2001. For 1999, however, Pray *et al.* (2001) report that Chinese suppliers just covered their costs, while Monsanto/Delta and Pine Land received less than US$2 million. Besides greater formal competition in the seed sector in China, farmers also save and re-plant Bt seed. Saved seed thus competes with new seed, exerting

downward pressure on rents. Bt cotton acreage has roughly tripled in China from 1999 to 2001 (Huang *et al.*, 2002b), so it would be interesting to include estimates of seed sale rents captured in future analysis.

The world economic surplus from Bt cotton adoption in the USA and China was US$836 million in 2001. Chinese producers capture 51% of this gain with a US$428 million increase in producer surplus, while Chinese consumers capture 20% of world economic surplus with a US$167 million increase in consumer surplus. The rest of the world captured 8% of the gain, with consumer gains slightly exceeding producer losses. Losses to producers in the rest of the world accrue because of the falling world price of cotton. US producers captured US$179 million and consumers US$48 million.

US commodity programme payments shelter US producers from the impact of the falling world price, but at a budgetary cost. Under joint

Table 16.2. World welfare effects (US$ millions) of Bt cotton adoption in the USA and China, 2001.

	Bt cotton adoption in		
	USA only	China only	Both the USA and China
Rest of the world			
Change in consumer surplus	217	201	418
Change in producer surplus	−181	−168	−349
Change in welfare	36	33	69
China			
Change in consumer surplus	87	81	167
Change in producer surplus	−84	514	428
Change in welfare	3	595	595
USA			
Change in consumer surplus	25	23	48
Change in producer surplus	164	14	179
Change in government payments	134	63	198
Seed supplier US profits	143	0	143
Change in welfare	198	−26	172
Global welfare	237	602	836

US–Chinese adoption, US producer surplus would have declined if not for commodity programme payments of US$198 million. This result is similar to one obtained by Sobelevsky *et al.* (2002) in their analysis of global adoption of transgenic soybeans. They found that US producers gained from global adoption of biotechnology with loan deficiency payments in place, but, in general, not when they were absent. When the USA adopts Bt cotton alone, programme payments account for 82% of producer surplus gains.

The USA captured 21% of the increase in world economic surplus, with 17 of this 21% going to seed suppliers as profit. Relative gains in China were larger, in part, because China was starting from a base of less effective pest control. Bt cotton adoption led to greater yield increases and greater reductions in insect control costs in China than in the USA.

Table 16.2 also highlights the consequences of falling behind technologically. If Bt cotton were adopted in China but not the USA, US welfare would fall by US$26 million. If Bt cotton were adopted in the USA, but not China, then welfare in China would only increase by US$3 million, instead

of US$595 million with combined adoption. With only US adoption, producer surplus in China falls by US$84 million.

Under joint adoption, producers in the rest of the world are worse off by US$349 million. As noted earlier, producers in the rest of the world were almost entirely non-adopters of Bt cotton in 2001 (producers in Australia are a notable exception). Unlike Bt cotton adopters in China and the USA, producers in the rest of the world do not benefit from higher yields and lower per pound production costs. They are only negatively affected by US–Chinese adoption through the falling world price of cotton.

The costs of falling behind technologically have implications for pest resistance management. The difference in US producer surplus when both the USA and China adopt Bt cotton and when only China adopts may be interpreted as the cost to US producers of resistance to Bt developing in the USA. A similar comparison could be made to estimate the cost of Bt resistance developing in China.

In simulation exercises, Edwards and Freebairn (1984) found that technological spillovers across regions reduce the gains from technological change in net exporting regions, while increasing the gains to net importing regions. Our results are consistent with these earlier findings. In the case of cotton, the USA is a net exporter, while the rest of the world and China are net importers. US welfare is highest when it adopts alone, while welfare is highest for China and the rest of the world when there is combined adoption.

Concluding Remarks

This chapter presented simulation results from a three-region, output price endogenous model of the world cotton market to evaluate the global impacts of Bt cotton adoption in the USA and China in 2001. Bt cotton reduced insecticide use and per pound production costs in both countries. Higher yields and production contributed to a 1.4 cent per pound reduction in the world price of cotton. Net global benefits were US$836 million. China captured 71% of this benefit, the USA captured 21% and the rest of the world captured the remainder. Cotton purchasers from the rest of the world benefited from lower cotton prices, while returns to their producers fell.

The results provide an indication of the costs of *not* adopting Bt cotton. Both the USA and China were worse off under scenarios when they did not adopt Bt cotton. Producers in the rest of the world lost US$349 million. Bt cotton adoption rates were negligible in the rest of the world in 2001 and assumed to be zero in the model. In the simulations, producers in the rest of the world lose from falling prices, but do not benefit from lower insecticide use or higher yields from adopting Bt cotton. In contrast, producers in China gained more than any other group examined in the simulations, gaining US$595 million.

An interesting extension of this present work would be to expand the regions modelled to the USA, China, India and the rest of the world. Bt cotton was first approved for use in India in 2002. Combined, the USA, China and India account for about 55% of world cotton production. Field trial results suggest that productivity gains in India from Bt cotton adoption could be substantial (Qaim and Zilberman, 2003). Some projections of adoption rates in China predict that Bt cotton's share of total cotton acreage will rise from 30% in 2001 to 79% by 2005 (Huang *et al.*, 2002b). Large-scale adoption in India and China can have

important implications for future world cotton prices and trade patterns.

References

Carpenter, J. and Gianessi, L. (2001) *Agricultural Biotechnology: Updated Benefits Estimates.* Case National Center for Food and Agricultural Policy, Washington, DC.

Cotton Research and Development Council (2002) *Annual Report 2001–2.* Narrabri NSW, Australia.

Deepak, M.S., Spreen, T.H. and Van Sickle, J. (1996) An analysis of the impact of a ban of methyl bromide on the U.S. fresh vegetable market. *Journal of Agricultural and Applied Economics* 28, 433–443.

Doyle, B., Reeve, I. and Barclay, E. (2002) *The Performance of Ingard Cotton in Australia During the 2000/2001 Season.* Institute for Rural Futures. University of New England, Armidale, New South Wales.

Edge, J.M., Benedict, J.H., Carroll, J.P. and Reding, H.K. (2001) Bollgard cotton: an assessment of global economic, environmental and social benefits. *Journal of Cotton Science* 5, 121–136.

Edwards, G.W. and Freebairn, J.W. (1984) The gains from research into tradable commodities. *American Journal of Agricultural Economics* 66, 41–49.

Frisvold, G. and Tronstad, R. (2002) Economic impacts of Bt cotton adoption in the United States. In: Kalaitzandonakes, N. (ed.) *The Economic and Environmental Impacts of Agbiotech: a Global Perspective.* Kluwer-Plenum, Norwell, Massachusetts, pp. 261–286.

Frisvold, G., Tronstad, R. and Mortensen, J. (2000) Bt cotton adoption: regional differences in producer costs and returns. *Proceedings of the Beltwide Cotton Conferences* 1, 337–340.

Gianessi, L.P., Silvers, C.S., Sankula, S. and Carpenter, J.E. (2002) *Plant Biotechnology: Current and Potential Impact for Improving Pest Management in U.S. Agriculture, An Analysis of 40 Case Studies.* National Center for Food and Agricultural Policy, Washington, DC.

Huang, J., Hu, R., Rozelle, S., Qiao, F. and Pray, C.E. (2002a) Transgenic varieties and productivity of smallholder cotton farmers in China. *Australian Journal of Agricultural and Resource Economics* 46, 367–87.

Huang, J., Hu, R., van, H. and van Tongeren, F. (2002b) *Biotechnology Boosts to Crop Productivity in China and its Impact on Global Trade.* Chinese Center for Agricultural Policy, Working Paper 02-E7.

Huang, J., Rozelle, S., Pray, C. and Wang, Q. (2002c) Plant biotechnology in China. *Science* 295, 674–677.

Isengildina, O., Hudson, D. and Herndon, C.W. (2000) The export elasticity of demand revisited: implications of changing markets. *Proceedings of the Beltwide Cotton Conferences* 1, 265–269.

Ismael, Y., Bennett, R. and Morse, S. (2002) Benefits from Bt cotton use by smallholder farmers in South Africa. *AgBioForum* 5, 1–5.

James, C. (2001) *Preview: Global Review of Commercialized Transgenic Crops: 2001.* ISAAA Briefs No. 24. ISAAA: Ithaca, New York.

Lichtenberg, E., Parker, D. and Zilberman, D. (1988). Marginal analysis of welfare costs of environmental policies: the case of pesticide regulation. *American Journal of Agricultural Economics* 70, 867–874.

Marra, M.C. (2001) The farm level impacts of transgenic crops: a critical review of the evidence. In: Pardey, P.G. (ed.) *The Future of Food: Biotechnology Markets in an International Setting.* Johns Hopkins Press and International Food Policy Research Institute, Baltimore, Maryland, pp. 155–184.

Meyer, L.A. (1999) An economic analysis of U.S. total fiber demand and cotton mill demand. *Cotton and Wool Situation and Outlook Yearbook.* US Department of Agriculture, Economic Research Service, Washington, DC, pp. 23–28.

Moschini, G. and Lapan, H. (1997) Intellectual property rights and the welfare effects of agricultural R&D. *American Journal of Agricultural Economics* 79, 1229–1242.

Pray, C., Ma, D., Huang, J. and Qiao, F. (2001) Impact of Bt cotton in China. *World Development* 29, 813–825.

Pray, C., Huang, J. Hu, R. and Rozelle, S. (2002) Five years of Bt cotton in China – the benefits continue. *Plant Journal* 34, 423–430.

Price, G.K., Lin,W. and Falck-Zepeda, J. (2001) *The Distribution of Benefits Resulting from Biotechnology Adoption.* Paper presented at the American Agricultural Economics Association Annual Meeting, Chicago, Illinois, August 5–8, 2001.

Qaim, M. and de Janvry, A. (2002) *Bt Cotton in Argentina: Analyzing Adoption and Farmers' Willingness to Pay.* Paper presented at the American Agricultural Economics Association Annual Meeting, Long Beach, California, July 28–31, 2002.

Qaim, M. and Zilberman, D. (2003) Yield effects of genetically modified crops in developing countries. *Science* 299, 900–902.

Sobolevsky, A., Moschini, G. and Lapan, H. (2002) *Genetically Modified Crop Innovations and Product Differentiation: Trade and Welfare Effects in the Soybean Complex.* Working Paper 02-WP 319. Center for Agricultural and Rural Development Iowa State University.

Sullivan, J., Roningen, V. and Waino, J. (1989) *A Database for Trade Liberalization Studies.* Staff Report AGES 89-12. Economic Research Service, US Department of Agriculture.

Sunding, D.L. (1996) Measuring the marginal cost of nonuniform environmental regulations. *American Journal of Agricultural Economics* 78, 1098–1107.

Traxler, G., Godoy-Avila, S., Falck-Zepeda, J. and Espinoza-Arellano, J. (2002) Transgenic cotton in Mexico: economic and environmental impacts. In: Kalaitzandonakes, N. (ed.) *The Economic and Environmental Impacts of Agbiotech: a Global Perspective.* Kluwer-Plenum, Norwell, Massachusetts, pp. 183–202.

Williams, M.R. (various years). Cotton insect losses. *Proceedings of the Beltwide Cotton Conferences.*

Index

Page numbers in **bold** refer to illustrations and tables

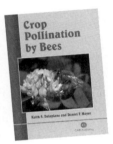
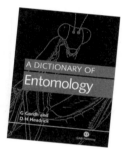
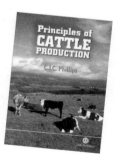